OXFORD STUDIES IN PROBABILITY

OXFORD STUDIES IN PROBABILITY

Frank B. Knight: *Foundations of the prediction process*
A. D. Barbour, Lars Holst, and Svante Janson: *Poisson approximation*
J. F. C. Kingman: *Poisson processes*

OXFORD STUDIES IN PROBABILITY · 3

Poisson Processes

J. F. C. KINGMAN

University of Bristol

CLARENDON PRESS · OXFORD

1993

Oxford University Press, Walton Street, Oxford OX2 6DP
Oxford New York Toronto
Delhi Bombay Calcutta Madras Karachi
Kuala Lumpur Singapore Hong Kong Tokyo
Nairobi Dar es Salaam Cape Town
Melbourne Auckland Madrid
and associated companies in
Berlin Ibadan

Oxford is a trade mark of Oxford University Press

Published in the United States
by Oxford University Press Inc., New York

A catalogue record for this book is available from the British Library

Library of Congress Cataloging in Publication Data
Kingman, J. F. C. (John Frank Charles)
Poisson processes / J.F.C. Kingman.
p. cm. — (Oxford studies in probability ; 3)
Includes bibliographical references and index.
1. Poisson processes. I. Title. II. Series.
QA274.42.K56 1993 519.2'3—dc20 92-25532

ISBN 0–19–853693–3

Typeset by Integral Typesetting, Gt. Yarmouth, Norfolk
Printed in Great Britain on acid-free paper by Bookcraft Ltd., Midsomer Norton, Avon

Preface

In the theory of random processes there are two that are fundamental, and occur over and over again, often in surprising ways. There is a real sense in which the deepest results are concerned with their interplay. One, the Bachelier–Wiener model of Brownian motion, has been the subject of many books. The other, the Poisson process, seems at first sight humbler and less worthy of study in its own right. Nearly every book mentions it, but most hurry past to more general point processes or Markov chains.

This comparative neglect is ill judged, and stems from a lack of perception of the real importance of the Poisson process. This distortion comes in turn partly from a restriction to one dimension, and the theory becomes more natural and more powerful in a more general context.

This book attempts, in a small compass, to do something to redress the balance. It records a long fascination with the beauty and wide application of Poisson processes in one or more dimensions, a fascination deepened by discussions over the years with many friends. I have used their ideas without acknowledgement, and I hope they will enjoy reading a book that would not have been written without their unwitting help. Particular thanks are due to David Aldous, the late Rollo Davidson, Warren Ewens, Frank Kelly, David Kendall, Dennis Lindley, Bob Loynes, the late Pat Moran, Geoff Watterson, Peter Whittle, and David Williams.

I have not tried to document the historical development of the theory, and the references are for necessary results or as suggestions for further reading. An accurate history would need to record the frequent rediscovery of the basic properties of the Poisson process, which took many decades to become widely known. Even now, there are widespread misconceptions, which it is hoped that this account will help to dispel.

The writer on probability always has a dilemma, since the mathematical basis of the theory is the arcane subject of abstract measure theory, which is unattractive to many potential readers. I have tried to resolve the problem by isolating measure-theoretic argument in a few proofs where it is essential, but presenting most of the development in a less pedantic style. The reader who wishes can safely take the measure theory for granted, but should nevertheless appreciate that a few obvious results need non-trivial proof.

There is some standard notation that recurs throughout, such as $\mathbb{P}$ and $\mathbb{E}$ for probability and expectation, $\mathbb{R}^d$ for Euclidean space of dimension d, and $\#$ for the number of points in a set. Script capitals are used for standard distributions (most commonly $\mathscr{P}$ for the Poisson distribution). Words like 'positive' and 'increasing' are used in the weak sense unless qualified by the adverb 'strictly'.

Bristol J.F.C.K.
April 1992

Contents

1

Stochastic models for random sets of points

1.1 Poisson models

Figure 1.1 might represent the positions of visible stars in a patch of the sky, the positions (seen from above) of trees in an area of woodland, or those of certain archaeological sites. There is no clear pattern, the points are haphazardly distributed with no obvious regularity or trends in density. Similar situations can arise in three dimensions or in one, or in more complicated geometries. If we look at the whole sky, for example, the stars appear to be distributed over the surface of a sphere. Poisson processes are models of such phenomena, which use the theory of probability to describe this sort of highly random behaviour. Their characteristic feature is a property of statistical independence.

To explain this, suppose that some 'test sets' $A_1, A_2, \ldots$ have been set down in the space of Fig. 1.1, their shapes and positions being determined without reference to the random points (Fig. 1.2). Denote by $N(A)$ the number of points falling in a set A. Then the numbers $N(A_j)$ are integer-valued random variables. *We assume that, as long as the sets $A_1, A_2, \ldots,$ do not overlap, these random variables are statistically independent.*

In practice of course this assumption is satisfied, if at all, only approximately.

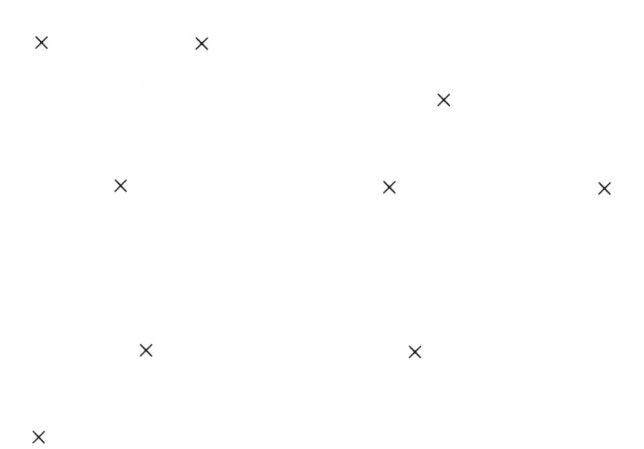

Fig. 1.1 A two-dimensional Poisson process.

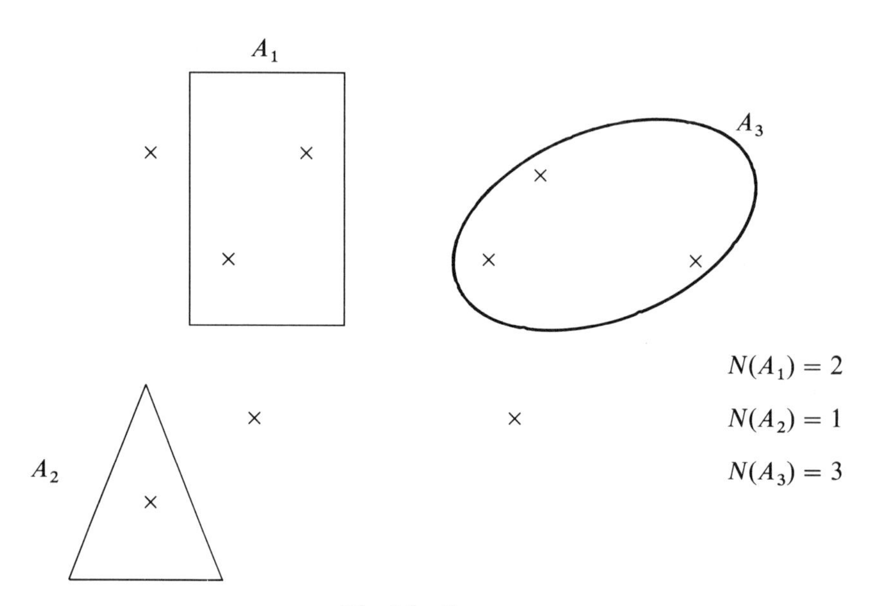

Fig. 1.2 Test sets.

There may be subtle effects which allow the presence of points in one region to influence those in another. There may be limitations which prevent two points from being too close together. There may be unexpected geometric patterns or regularities ('ley lines' for instance in archaeology). A Poisson model is usually the simplest and, in a sense, the most random way in which to describe any particular phenomenon.

The name *Poisson* comes from the probability distribution of the same name, which plays a central role in the theory. This in turn is a limiting case of the binomial distribution, and as such goes back to the work of S.-D. Poisson (1837) himself. Thus if a coin is tossed n times, if the tosses are independent and have the same probability p of showing a head, the probability that there are r heads and $n - r$ tails is

$$b(n, p; r) = \binom{n}{r} p^r (1 - p)^{n-r} \tag{1.1}$$

If $\mu = np$, the average number of heads, is held constant while n becomes large and p small, the expression (1) has a limit

$$\pi_r(\mu) = \lim_{n \to \infty} b\left(n, \frac{\mu}{n}; r\right)$$

which is easily seen to be

$$\pi_r(\mu) = \mu^r \, e^{-\mu}/r! \tag{1.2}$$

for $r \geqslant 0$, and this is called the *Poisson distribution with parameter* μ, which we denote by the symbol $\mathscr{P}(\mu)$. Similarly the binomial distribution (1.1) is denoted by $\mathscr{B}(n, p)$.

The Poisson process, as distinct from the Poisson distribution, first arose in one dimension, representing time. The one-dimensional version of Fig. 1.1 would look like Fig. 1.3, where the points could now represent a random series of events occurring in time. Physicists use Poisson processes to describe the emission of radioactive particles, while teletraffic engineers have used them to account for the arrival times of telephone calls at an exchange.

Fig. 1.3 A one-dimensional Poisson process.

This is natural if we regard the events as emanating from different atoms (in the first case) or subscribers (in the second) which are acting independently of each other. It is unlikely that any one atom will contribute an event in a particular interval of time, but there are many such and it is likely that a few will do so. This is exactly the situation in which the binomial probability (1.1) is well approximated by its Poisson limit (1.2).

Thus we shall assume that the counts $N(A)$ are not merely independent for disjoint A but also that each has the Poisson distribution $\mathscr{P}(\mu)$ for a suitable choice of μ. These assumptions are compatible in a sense that will be made precise, and indeed the assumption of independence leads almost inevitably to the Poisson distribution (Section 1.4).

Our object is to develop a theory which is relevant to a wide variety of applications, and we shall therefore make only minimal assumptions about the space ($\mathbb{R}^2$ in Fig. 1.1, $\mathbb{R}$ in Fig. 1.3) in which the random points sit. It might be thought that the one-dimensional case is simpler, but in fact the essential simplicity of the general theory can be obscured by special features of the real line, notably the possibility of numbering the random points in order. The reader would be well advised, throughout the general theory, to think in terms of two-dimensional pictures like Fig. 1.1 rather than in one dimension as in Fig. 1.3.

1.2 The Poisson distribution

A random variable X has the Poisson distribution $\mathscr{P}(\mu)$ if its possible values are positive integers and if

$$\mathbb{P}\{X = n\} = \pi_n(\mu) = \mu^n \, e^{-\mu}/n! \tag{1.3}$$

for $n \geqslant 0$. Here μ can take any value in $\mu > 0$, and the identity

$$\sum_{n=0}^{\infty} n\pi_n(\mu) = \mu \tag{1.4}$$

shows that μ is the mean of the distribution $\mathscr{P}(\mu)$:

$$\mathbb{E}(X) = \mu. \tag{1.5}$$

It is sometimes useful to extend the definition of $\mathscr{P}(\mu)$ to include the boundary cases 0 and ∞. Thus $\mathscr{P}(0)$ will mean the distribution concentrated at 0:

$$\mathbb{P}\{X = 0\} = 1, \tag{1.6}$$

and $\mathscr{P}(\infty)$ the distribution concentrated at $+\infty$:

$$\mathbb{P}\{X = +\infty\} = 1. \tag{1.7}$$

If z is any real or complex number in $|z| \leqslant 1$, the random variable z^X is bounded, so that

$$\begin{aligned}\mathbb{E}(z^X) &= \sum_{n=0}^{\infty} \pi_n(\mu) z^n \\ &= \mathrm{e}^{-\mu} \sum_{n=0}^{\infty} \frac{(\mu z)^n}{n!} = \mathrm{e}^{-\mu(1-z)},\end{aligned} \tag{1.8}$$

and this holds for $0 \leqslant \mu < \infty$. Identities for the moments of X can then be derived by differentiation and setting $z = 1$ (or by direct summation):

$$\begin{aligned}\mathbb{E}\{X\} &= \mu, \\ \mathbb{E}\{X(X-1)\} &= \mu^2, \\ \mathbb{E}\{X(X-1)(X-2)\} &= \mu^3, \ldots\end{aligned}$$

from which we can deduce that

$$\mathbb{E}\{X^2\} = \mu + \mu^2, \qquad \mathrm{var}\{X\} = \mu, \tag{1.9}$$

that

$$\mathbb{E}\{X^3\} = \mu + 3\mu^2 + \mu^3, \tag{1.10}$$

and so on.

Differentiating π_n with respect to μ, we have

$$\frac{\mathrm{d}\pi_n}{\mathrm{d}\mu} = \pi_{n-1} - \pi_n \tag{1.11}$$

with the convention that $\pi_{-1} = 0$. Thus

$$\frac{\mathrm{d}}{\mathrm{d}\mu} \sum_{k=0}^{n} \pi_k = -\pi_n.$$

Integrating from 0 to μ gives the useful identity

$$\sum_{k=0}^{n} \pi_k(\mu) = 1 - \int_0^{\mu} \pi_n(\lambda)\, \mathrm{d}\lambda. \tag{1.12}$$

The most important feature of the Poisson distribution is its additivity. If X and Y are *independent* random variables, with Poisson distributions $\mathscr{P}(\lambda)$ and $\mathscr{P}(\mu)$ respectively, then for $r, s \geqslant 0$,

$$\mathbb{P}\{X = r, Y = s\} = \mathbb{P}\{X = r\}\mathbb{P}\{Y = s\} = \frac{\lambda^r \mathrm{e}^{-\lambda}}{r!} \frac{\mu^s \mathrm{e}^{-\mu}}{s!}. \tag{1.13}$$

We can find the distribution of the random variable $X + Y$ by summing (1.13) over values of r and s with fixed $r + s$;

$$\begin{aligned}
\mathbb{P}\{X + Y = n\} &= \sum_{r=0}^{n} \mathbb{P}\{X = r, Y = n - r\} \\
&= \sum_{r=0}^{n} \frac{\lambda^r \mathrm{e}^{-\lambda}}{r!} \frac{\mu^{n-r} \mathrm{e}^{-\mu}}{(n-r)!} \\
&= \frac{\mathrm{e}^{-(\lambda+\mu)}}{n!} \sum_{r=0}^{n} \binom{n}{r} \lambda^r \mu^{n-r} \\
&= \frac{(\lambda + \mu)^n \mathrm{e}^{-(\lambda+\mu)}}{n!}.
\end{aligned}$$

Thus $X + Y$ has distribution $\mathscr{P}(\lambda + \mu)$. This extends by induction to the sum of any finite number of independent random variables. We shall however need an even more powerful result, which covers infinite sums and gives conditions for convergence.

Countable Additivity Theorem *Let X_j ($j = 1, 2, \ldots$) be independent random variables, and assume that X_j has the distribution $\mathscr{P}(\mu_j)$ for each j. If*

$$\sigma = \sum_{j=1}^{\infty} \mu_j \tag{1.14}$$

converges, then

$$S = \sum_{j=1}^{\infty} X_j \tag{1.15}$$

converges with probability 1, *and S has distribution $\mathscr{P}(\sigma)$. If on the other hand* (1.14) *diverges, then S diverges with probability* 1.

Proof By induction on n,

$$S_n = \sum_{j=1}^{n} X_j$$

has distribution $\mathscr{P}(\sigma_n)$, where

$$\sigma_n = \sum_{j=1}^{n} \mu_j.$$

Thus, for any r,

$$\mathbb{P}\{S_n \leqslant r\} = \sum_{k=0}^{r} \pi_k(\sigma_n).$$

The events $\{S_n \leqslant r\}$ decrease as n increases for fixed r, and their intersection is $\{S \leqslant r\}$. Thus

$$\begin{aligned}\mathbb{P}\{S \leqslant r\} &= \lim_{n\to\infty} \mathbb{P}\{S_n \leqslant r\} \\ &= \lim_{n\to\infty} \sum_{k=0}^{r} \pi_k(\sigma_n).\end{aligned}$$

If σ_n converges to a finite limit σ, the continuity of π_k shows that

$$\mathbb{P}\{S \leqslant r\} = \sum_{k=0}^{r} \pi_k(\sigma)$$

so that

$$\mathbb{P}\{S = r\} = \pi_r(\sigma)$$

and so S is finite with distribution $\mathscr{P}(\sigma)$. But if $\sigma_n \to \infty$,

$$\sum_{k=0}^{r} \pi_k(\sigma_n) = \mathrm{e}^{-\sigma_n} \sum_{k=0}^{r} \frac{\sigma_n^k}{k!} \to 0,$$

so that

$$\mathbb{P}\{S > r\} = 1.$$

Since this holds for all r, S diverges with probability 1, and the proof is complete.

Note that this theorem shows the naturalness of the conventional definitions of $\mathscr{P}(0)$ and $\mathscr{P}(\infty)$. With this convention, it states that, if independent random variables X_j have distributions $\mathscr{P}(\mu_j)$ respectively, then their sum has distribution $\mathscr{P}(\sum \mu_j)$. This is true whether there are finitely or infinitely many of them, and even if some of the μ_j are zero or infinite or both.

Let $X_1, X_2, \ldots, X_n$ be independent, and let X_j have distribution $\mathscr{P}(\mu_j)$ for all j. Then $S = X_1 + \cdots + X_n$ has distribution $\mathscr{P}(\sigma)$, *with* $\sigma = \sum \mu_j$; and so, whenever $r_1 + r_2 + \cdots + r_n = s$,

$$\begin{aligned}\mathbb{P}\{X_1 = r_1, X_2 = r_2, \ldots, X_n = r_n | S = s\} &= \prod_{j=1}^{n} \frac{\mu_j^{r_j} e^{-\mu_j}}{r_j!} \Big/ \frac{\sigma^s e^{-\sigma}}{s!} \\ &= \frac{s!}{r_1! r_2! \cdots r_n!} \left(\frac{\mu_1}{\sigma}\right)^{r_1} \left(\frac{\mu_2}{\sigma}\right)^{r_2} \cdots \left(\frac{\mu_n}{\sigma}\right)^{r_n}. \end{aligned} \tag{1.16}$$

These are the probabilities of the *multinomial distribution* $\mathcal{M}(s; p_1, \ldots, p_n)$, where $p_j = \mu_j/\sigma$.

In the special case $n = 2$, this reduces to the result that, if X and Y are independent Poisson variables then, given that $X + Y = n$, the conditional distribution of X is $\mathcal{B}(n, p)$, where

$$p = \frac{\mathbb{E}(X)}{\mathbb{E}(X) + \mathbb{E}(Y)}.$$

Thus the binomial and multinomial distributions can be obtained by conditioning from appropriate Poisson distributions.

There is a partial converse which is sometimes useful. Suppose that N is Poisson with mean μ, and that M, given N, has distribution $\mathcal{B}(N, p)$ for a fixed constant p. By this we mean that, for $0 \leqslant t \leqslant s$,

$$\mathbb{P}\{M = t | N = s\} = b(s, p; t). \tag{1.17}$$

Then, for $m, k \geqslant 0$,

$$\begin{aligned}\mathbb{P}\{M = m, N - M = k\} &= \mathbb{P}\{N = m + k\}\mathbb{P}\{M = m | N = m + k\} \\ &= \frac{e^{-\mu} \mu^{m+k}}{(m+k)!} \binom{m+k}{m} p^m (1-p)^k \\ &= \frac{e^{-\mu p}(\mu p)^m}{m!} \frac{e^{-\mu(1-p)}(\mu(1-p))^k}{k!}.\end{aligned}$$

Thus M and $N - M$ are independent Poisson variables, with means μp and $\mu(1 - p)$.

1.3 Probability spaces for Poisson processes

This section is for the purist, who rightly demands that random variables and probabilities be defined by reference to a proper probability space, and that the measure-theoretic niceties be observed. It may be omitted without loss by the more casual reader.

According to the accepted basis of probability theory (Billingsley 1979, Breiman 1968, Chung 1968, Doob 1953, Kingman and Taylor 1966, Rényi 1970), we should first define a triple $(\Omega, \mathscr{F}, \mathbb{P})$ where Ω is a set (of 'elementary outcomes'), $\mathscr{F}$ a σ-field of subsets of Ω ('events') and $\mathbb{P}$ a probability measure assigning a number in $[0, 1]$ to every event in $\mathscr{F}$. A real-valued random variable is a function X from Ω into $\mathbb{R}$ which is measurable in the sense that $\{\omega; X(\omega) \leqslant x\}$ belongs to $\mathscr{F}$ for every x. A complex random variable is likewise a measurable function from Ω into $\mathbb{C}$, a random n-vector a function from Ω into $\mathbb{R}^n$, and a random elephant is a function from Ω into a suitable space of elephants. In each case there is a measurability condition to ensure that interesting subsets of Ω belong to $\mathscr{F}$ and so have probabilities.

Random arrays of the type represented by Fig. 1.1 are random finite or perhaps countably infinite subsets of the underlying space S in which the points sit, which is called the *state space*. Thus a Poisson process, with state space S, defined on a probability space $(\Omega, \mathscr{F}, \mathbb{P})$, is a function Π from Ω into the set S^∞ of all countable subsets of S.

We now need to distinguish, among the subsets of S, some which we wish to use as 'test sets' in the manner of Fig. 1.2. If A is such a set, the number of points of Π in A is

$$N(A) = \#\{\Pi(\omega) \cap A\}, \tag{1.18}$$

where the symbol $\#$ denotes the number of points in the set following it, 0 if the set is empty, and ∞ if the set is infinite. Because the right-hand side depends on ω, each $N(A)$ is a function

$$N(A)\colon \Omega \to \{0, 1, 2, \ldots, \infty\} \tag{1.19}$$

and we require this function to be measurable for each test set A. That is, we suppose that, for each test set A and each n,

$$\{\omega; N(A) = n\} \in \mathscr{F}. \tag{1.20}$$

Then $N(A)$ are then random variables, and we can impose conditions on their distributions and joint distributions. It is rarely necessary to be too punctilious about the subsets of S which are considered as test sets. This is because we can always build up more complicated sets out of simple ones by operations for which the behaviour of N is obvious.

For example, if S is the real line $\mathbb{R}$ (Fig. 1.3) it is sufficient that (1.20) be true whenever A is an open interval (a, b). Every open set G is a countable union of disjoint open intervals A_j, and then $N(G) = \sum N(A_j)$ is a random variable if the $N(A_j)$ are. Every closed set F is the intersection of a decreasing sequence of open sets G_i, and $N(F)$ is the limit of the decreasing sequence $N(G_i)$. In this way we can show that $N(A)$ is a random variable for any set A we are likely to encounter (indeed for any Borel set).

The same is true if $S = \mathbb{R}^2$ (Fig. 1.1), so long as (1.20) holds whenever A

is an open rectangle. There are corresponding results for $\mathbb{R}^d$ and for more complicated geometries, but we shall not pursue them because they are unnecessary for the approach we take to the existence problem in the next chapter. Rather similar questions will however become urgent in Section 3.4, in a slightly different context.

In Section 2.1 a Poisson process will be formally defined as a random countable subset Π of S for which the random variables $N(A)$ counting the number of points of Π in the test sets A have two properties. One is the independence property described in Section 1.1. The other is that $N(A)$ has the Poisson distribution $\mathscr{P}(\mu)$, where the parameter μ depends on A. The dependence of μ on A is not arbitrary, and in fact μ must be a non-atomic measure on S.

Conversely, suppose that μ is a non-atomic measure on S. Can we be sure that there is a Poisson process for which $N(A)$ has distribution $\mathscr{P}(\mu(A))$ for all A? General theorems on the existence of random processes show that there are families of random variables $N(A)$ with the right joint distributions, but it is much more difficult (Kendall 1974) to show that these come from a random set Π through (1.18).

It is much easier to give an explicit construction, and this is done in Section 2.5 under very weak conditions on S and μ. These conditions are easily satisfied in all known applications, and the result is a theory of great generality and wide applicability. Before describing this, however, we show in the next section that it is only the Poisson distribution which can possibly give rise to such a simple theory.

1.4 The inevitability of the Poisson distribution

Think of a test set A_t which is in two dimensions a circle of radius t (though it could equally be a square of side t). Write

$$p_n(t) = \mathbb{P}\{N(A_t) = n\}, \tag{1.21}$$

$$q_n(t) = \mathbb{P}\{N(A_t) \leqslant n\}. \tag{1.22}$$

Because $N(A_t)$ increases with t, q_n decreases, so that q_n and p_n have only jump discontinuities and are differentiable almost everywhere. We give a heuristic derivation of a differential equation for p_n.

The random variable $N(A_t)$ jumps from n to $n + 1$ when its enlarging boundary crosses one of the random points. The probability that this occurs between t and $t + h$ is the probability that there is a point in the ring between A_t and A_{t+h}, which we suppose small when h is small. If we ignore the probability of two or more points in this ring, this probability (given the number in A_t) is the same as $\mu(t + h) - \mu(t)$, where

$$\mu(t) = \mathbb{E}\{N(A_t)\}. \tag{1.23}$$

If this is independent of the number of points in A_t, then for small h,

$$q_n(t) - q_n(t+h) = p_n(t)\{\mu(t+h) - \mu(t)\}.$$

Letting $h \to 0$,

$$-\frac{\mathrm{d}q_n}{\mathrm{d}t} = p_n \frac{\mathrm{d}\mu}{\mathrm{d}t}. \tag{1.24}$$

Since $p_n = q_n - q_{n-1}$, this means that

$$\frac{\mathrm{d}p_0}{\mathrm{d}t} = -p_0 \frac{\mathrm{d}\mu}{\mathrm{d}t}, \qquad \frac{\mathrm{d}p_n}{\mathrm{d}t} = (p_{n-1} - p_n)\frac{\mathrm{d}\mu}{\mathrm{d}t} \quad (n \geqslant 1). \tag{1.25}$$

The first of these equations shows that

$$\frac{\mathrm{d}}{\mathrm{d}t}(\log p_0 + \mu) = 0$$

and since $p_0(0) = 1$, $\mu(0) = 0$,

$$\log p_0 + \mu = 0$$

so that

$$p_0(t) = \mathrm{e}^{-\mu(t)} \tag{1.26}$$

for all t. The second equation in (1.25) can be written

$$\frac{\mathrm{d}}{\mathrm{d}t}(p_n \mathrm{e}^{\mu}) = p_{n-1} \mathrm{e}^{\mu} \frac{\mathrm{d}\mu}{\mathrm{d}t}$$

so that, since $p_n(0) = 0$ $(n \geqslant 1)$,

$$p_n(t) = \mathrm{e}^{-\mu(t)} \int_0^t p_{n-1}(s)\, \mathrm{e}^{\mu(s)} \frac{\mathrm{d}\mu}{\mathrm{d}s}\, \mathrm{d}s. \tag{1.27}$$

From this it follows by induction on n, starting at (1.26), that

$$p_n(t) = \mathrm{e}^{-\mu(t)}\, \mu(t)^n/n! \tag{1.28}$$

so that $N(A_t)$ has distribution $\mathscr{P}(\mu(t))$.

This argument has clearly relied on several implicit assumptions, but it does suggest that the Poisson distribution is an inevitable consequence of the 'complete randomness' inherent in our assumptions of independence. We shall return to this more systematically in Chapter 8.

2

Poisson processes in general spaces

2.1 Definition and basic properties

The state space S in which the points of a Poisson process sit is usually Euclidean space of some dimension d, or more generally (Chapter 7) a manifold which is locally equivalent to $\mathbb{R}^d$. The theory does not, however, make use of the special properties of Euclidean space, but only needs a reasonable family of subsets to be used as test sets to count the random points. That is, we need sets for which the count function

$$N(A) = \#\{\Pi \cap A\} \tag{2.1}$$

is a well-defined random variable.

The natural way to do this is to suppose that S is a measurable space. We assume, that is to say, that certain subsets of S are called *measurable*, and that the measurable sets form a σ-field in the sense that:

- the empty set is measurable
- the complement of a measurable set is measurable
- the union of countably many measurable sets is measurable.

We also need to ensure that there are enough measurable sets to distinguish individual points. This can be done by making the weak assumption that the diagonal

$$D = \{(x, y); x = y\} \tag{2.2}$$

is measurable in the product space $S \times S$. This automatically implies that every singleton set $\{x\}$ in S is measurable.

We shall make these assumptions without further comment in all that follows. When $S = \mathbb{R}^d$ the measurable sets are always taken to be the Borel sets, those of the smallest σ-field containing the open sets. The diagonal assumption is certainly satisfied, because D is closed in the product space $\mathbb{R}^{2d}$.

A *Poisson process* on S is then a random countable subset Π of S, such that

(i) for any disjoint measurable subsets $A_1, A_2, \ldots, A_n$ of S, the random variables $N(A_1), N(A_2), \ldots, N(A_n)$ are independent, and

(ii) $N(A)$ has the Poisson distribution $\mathscr{P}(\mu)$, where $\mu = \mu(A)$ lies in $0 \leqslant \mu \leqslant \infty$.

Thus if $\mu(A)$ is finite, $\Pi \cap A$ is with probability 1 a finite set, empty if $\mu(A) = 0$. If $\mu(A) = \infty$, $\Pi \cap A$ is countably infinite with probability 1.

By (1.5)

$$\mu(A) = \mathbb{E}\{N(A)\}. \tag{2.3}$$

If $A_1, A_2, \ldots$ are disjoint, and have union A,

$$N(A) = \sum_{n=1}^{\infty} N(A_n),$$

and taking expectations,

$$\mu(A) = \sum_{n=1}^{\infty} \mu(A_n). \tag{2.4}$$

Thus μ is a measure on S which we shall call the *mean measure* (there is no standard term) of the Poisson process Π.

If we know the mean measure we can write down, using (i) and (ii), all the joint distributions of the counts $N(A)$ for different sets A. To see this, suppose that $A_1, A_2, \ldots, A_n$ are measurable, but not necessarily disjoint. Consider all sets of the form

$$B = A_1^* \cap A_2^* \cap \cdots \cap A_n^*$$

where each A_j^* is either A_j or its complement. There are 2^n of these, and they are disjoint. If they are listed as $B_1, B_2, \ldots, B_{2^n}$ then

$$A_j = \bigcup_{i \in \gamma_j} B_i$$

where γ_j is a subset of $\{1, 2, \ldots, 2^n\}$. Hence

$$N(A) = \sum_{i \in \gamma_j} N(B_i) \tag{2.5}$$

and the $N(B_i)$ are independent, with known Poisson distributions $\mathscr{P}(\mu(B_i))$. From this can be written down the probability of any event defined in terms of the random variables $N(A_1), N(A_2), \ldots, N(A_n)$.

For example, if $n = 2$ the listing can be done so that

$$A_1 = B_1 \cup B_2, \qquad A_2 = B_1 \cup B_3$$

and then

$$N(A_1) = N(B_1) + N(B_2), \qquad N(A_2) = N(B_1) + N(B_3).$$

Suppose for instance that we require the covariance of $N(A_1)$ and $N(A_2)$. Then we compute

$$\begin{aligned}\mathbb{E}\{N(A_1)N(A_2)\} &= \mathbb{E}\{N(B_1)^2 + N(B_1)N(B_3) + N(B_2)N(B_1) + N(B_2)N(B_3)\}\\ &= \{\mu(B_1)^2 + \mu(B_1)\} + \mu(B_1)\mu(B_3) + \mu(B_2)\mu(B_1) + \mu(B_2)\mu(B_3)\\ &= \mu(A_1)\mu(A_2) + \mu(B_1).\end{aligned}$$

Since $B_1 = A_1 \cap A_2$, this shows that

$$\operatorname{cov}\{N(A_1), N(A_2)\} = \mu(A_1 \cap A_2). \tag{2.6}$$

Not every measure can be a mean measure. Suppose that the measure μ on S has an atom at x, so that μ attaches non-zero measure to the singleton $\{x\}$; $m = \mu(\{x\}) > 0$. Then (ii) with $A = \{x\}$ shows that a Poisson process with mean measure μ would be such that

$$\mathbb{P}\{N(\{x\}) \geqslant 2\} = 1 - \mathrm{e}^{-m} - m\,\mathrm{e}^{-m} > 0.$$

This clearly contradicts (2.1), and shows that μ cannot be a mean measure.

Thus a mean measure must be non-atomic in the sense that

$$\mu(\{x\}) = 0 \qquad \text{for all } x \in S. \tag{2.7}$$

(Some authors have a stronger meaning for this term, but that will not concern us here.) It is a remarkable fact that this condition is almost sufficient for the existence of a Poisson process. We show in Section 2.5 that, under a very weak finiteness condition, every non-atomic measure is the mean measure of some Poisson process. Thus by varying the measure μ on S we get a wide variety of Poisson processes on a given state space S.

When $S = \mathbb{R}^d$, the mean measure is in most interesting cases given in terms of a *rate* or *intensity*. This is a positive measurable function λ on S, in terms of which μ is given by integrating λ with respect to d-dimensional measure:

$$\mu(A) = \int_A \lambda(x)\,\mathrm{d}x \tag{2.8}$$

(where $\mathrm{d}x$ stands for $\mathrm{d}x_1\,\mathrm{d}x_2 \cdots \mathrm{d}x_d$). The word rate is most often used when $d = 1$ and S is a time axis, while intensity tends to be used when S has a spatial interpretation. If λ is continuous at x, then (2.8) implies that for small neighbourhoods A of x,

$$\mu(A) \sim \lambda(x)|A|, \tag{2.9}$$

where $|A|$ denotes the measure (length if $d = 1$, area if $d = 2$, volume if $d = 3$, and so on) of A. Thus $\lambda(x)|A|$ is the approximate probability of a point of Π falling in the small set A, and is larger in regions where λ is large than in those where λ is small.

In the special case when λ is a constant, so that

$$\mu(A) = \lambda|A| \tag{2.10}$$

we speak of a *uniform* or *homogeneous* Poisson process. Such a random set has stochastic properties which are unchanged under translations or rotations of S, and is the only Poisson process (finite on bounded sets) with this property.

We have not assumed however that the integral in (2.8) necessarily converges, even on bounded sets. In most problems it does (but see Chapter 9 for a very important exception), so that $\mu(A)$, and therefore $N(A)$, is finite on bounded sets. Then with probability 1 the set Π is locally finite, and has no finite limit points.

Consider now the case $S = \mathbb{R}$, and suppose that $\mu(A)$ is finite on bounded sets A (but not necessarily given by a rate function λ). Then the measure μ is determined uniquely by its values on intervals $(a, b]$. Write

$$\begin{aligned} M(t) &= \mu(0, t] = \mathbb{E}\{N(0, t]\} \quad (t \geqslant 0), \\ M(t) &= -\mu(t, 0] = -\mathbb{E}\{N(t, 0]\} \quad (t < 0), \end{aligned} \tag{2.11}$$

so that M is an increasing function and

$$\mu(a, b] = M(b) - M(a) \quad (a < b). \tag{2.12}$$

Then μ is determined by the function M, and is then called the *Stieltjes measure* associated with the increasing function M.

The measure μ is non-atomic if and only if M is continuous. It may or may not be expressible as an indefinite integral

$$M(t) = \int_0^t \lambda(x)\, \mathrm{d}x \tag{2.13}$$

and if it is of this form then μ is given by (2.8). In particular, the uniform Poisson process of rate λ corresponds to

$$M(t) = \lambda t, \tag{2.14}$$

where λ is now a constant.

2.2 The Superposition Theorem

The Poisson process has a number of special properties which make its use, and the calculation of associated probabilities, often surprisingly simple. The three most important are the superposition property which is the subject of this section, the mapping property which is established in Section 2.3, and the various related selection, colouring and marking theorems which appear in full generality in Chapter 5. The superposition property is an almost immediate corollary of the Countable Additivity Theorem, except for an awkward technicality which is dealt with in the following lemma.

Disjointness Lemma *Let Π_1 and Π_2 be independent Poisson processes on S, and let A be a measurable set with $\mu_1(A)$ and $\mu_2(A)$ finite. Then Π_1 and Π_2 are disjoint with probability* 1 *on A:*

$$\mathbb{P}\{\Pi_1 \cap \Pi_2 \cap A = \varnothing\} = 1. \tag{2.15}$$

Proof Let A^f be the set of all finite subsets Λ of A. We make A^f a measurable space by giving it the smallest family of measurable sets which makes the function

$$\Lambda \mapsto \#(\Lambda \cap B) \tag{2.16}$$

measurable for all measurable $B \subseteq A$. Then $\Pi_1 \cap A$ is a random element of A^f, whose distribution is a probability measure P_1. Similarly $\Pi_2 \cap A$ has a distribution P_2 on A^f, and because Π_1 and Π_2 are independent the joint distribution of $\Pi_1 \cap A$ and $\Pi_2 \cap A$ is the product measure $P_1 \times P_2$ on $A^f \times A^f$.

Define the mapping η: $A^f \times A^f \to (A \times A)^f$ by

$$\eta(\Lambda_1, \Lambda_2) = \Lambda_1 \times \Lambda_2.$$

We first show that η is measurable, for which it suffices to show that $\eta^{-1}\{\Lambda; \#(\Lambda \cap C) = n\}$ is measurable in $A^f \times A^f$ for any measurable $C \subseteq A \times A$. This is true for $C = B_1 \times B_2$, because the assertion that

$$\#\{\eta(\Lambda_1, \Lambda_2) \cap (B_1 \times B_2)\} = n$$

is the assertion that, for some n_1, n_2 with $n_1 n_2 = n$,

$$\#(\Lambda_1 \cap B_1) = n_1, \qquad \#(\Lambda_2 \cap B_2) = n_2.$$

Moreover, the class of all such C is closed under disjoint unions and monotone limits, and so (see for instance Theorem 1.5 of Kingman and Taylor 1966) contains all measurable C in the product space $A \times A$.

In particular, since the diagonal D is measurable in $S \times S$, its restriction $D_A = D \cap (A \times A)$ is measurable in $A \times A$, so that

$$J = \eta^{-1}\{\Lambda; \#(\Lambda \cap D_A) = 0\}$$

is measurable in $A^f \times A^f$. The assertion (2.15) is equivalent to $(P_1 \times P_2)(J) = 1$ and by Fubini's theorem this is true if, for all Λ_1 outside a P_1-null set,

$$P_2\{\Lambda_2; (\Lambda_1, \Lambda_2) \in J\} = 1.$$

This, however, is just the assertion that

$$P_2\{N_2(\Lambda_1) = 0\} = 1,$$

and this is true since Λ_1 is finite, so that $\mu_2(\Lambda_1) = 0$ for all Λ_1. Thus the lemma is proved.

The result extends at once to sets A which are countable unions of sets on which μ_1 and μ_2 are both finite. In particular, it is true with $A = S$ if μ_1 and μ_2 are σ-finite. Some finiteness restriction is however essential, as the following example demonstrates.

Accepting for the moment that we have still to prove its existence, let Π

be a uniform Poisson process of rate $\lambda = 1$ on $\mathbb{R}^2$. If ϕ is the projection $\phi(x, y) = x$, the set

$$\phi(\Pi) = \{x; (x, y) \in \Pi\} \tag{2.17}$$

is a random countable subset of $\mathbb{R}$. Its counts have highly degenerate distributions; if A is a Borel set, then

$$\#\{\phi(\Pi) \cap A\} = \#\{\Pi \cap (A \times \mathbb{R})\}$$

which has a Poisson distribution with mean

$$\iint_{A \times \mathbb{R}} \mathrm{dx\, dy}$$

and this is 0 or ∞ according to whether $|A|$ is zero or non-zero. Hence the random variable $\#\{\phi(\Pi) \cap A\}$ is either 0 or ∞ with probability 1, and thus in either case has distribution $\mathscr{P}(\mu)$, with $\mu = 0$ or ∞. It follows from the definition that $\phi(\Pi)$ is a Poisson process on $\mathbb{R}$, with mean measure

$$\begin{aligned} \mu(A) &= 0 \quad \text{if } |A| = 0, \\ \mu(A) &= \infty \quad \text{if } |A| > 0. \end{aligned} \tag{2.18}$$

Now take $\Pi_1 = \Pi_2 = \phi(\Pi)$. The counts $N_1(A_1)$ and $N_2(A_2)$ of Π_1 and Π_2 are all degenerate, and therefore trivially independent of one another. Hence Π_1 and Π_2 are independent Poisson processes, but (2.15) is false when $|A| > 0$.

It might be objected that Π_1 and Π_2 are not 'really' independent, and only appear to be so because we choose to describe them in terms of their count processes. If we gave S^∞ a richer structure of measurable sets, we would be able to tell that Π_1 and Π_2 are just copies of the same random set $\phi(\Pi)$. The reader interested in pursuing such questions into some rather deep water should read Kendall (1974).

As it happens, these complications do not affect the immediate purpose of the Disjointness Lemma, which is to prove the following theorem.

Superposition Theorem *Let $\Pi_1, \Pi_2, \ldots$ be a countable collection of independent Poisson processes on S and let Π_n have mean measure μ_n for each n. Then their superposition*

$$\Pi = \bigcup_{n=1}^{\infty} \Pi_n \tag{2.19}$$

is a Poisson process with mean measure

$$\mu = \sum_{n=1}^{\infty} \mu_n. \tag{2.20}$$

Proof Let $N_n(A)$ be the number of points of Π_n in the measurable set A. If $\mu_n(A) < \infty$ for all n, the lemma shows that the random sets Π_n are disjoint on A, so that the number of points of Π in A is

$$N(A) = \sum_{n=1}^{\infty} N_n(A). \tag{2.21}$$

By the Countable Additivity Theorem $N(A)$ has distribution $\mathscr{P}(\mu)$, where $\mu = \mu(A)$ is given by (2.20). On the other hand, if $\mu_n(A) = \infty$ for some n, then $N_n(A) = N(A) = \infty$ and (2.21) holds trivially.

To prove the theorem we have only to prove that $N(A_1), N(A_2), \ldots, N(A_k)$ are independent if the set A_j are disjoint. This is clear because the double array of variables

$$N_n(A_j), \quad n = 1, 2, \ldots, \quad j = 1, 2, \ldots, k$$

are all independent, and $N(A_j)$ is defined in terms of a subset of these variables disjoint from those for other j.

This completes the proof, which for the sake of generality has been stated for countably infinite superpositions; it contains as an obvious corollary the corresponding result for finite superpositions.

To complete this section, we state another result that it is hardly worth stating. Its proof is just a matter of checking the definition, but it is used so often that it is proper to bring it into the open.

Restriction Theorem *Let Π be a Poisson process with mean measure μ on S, and let S_1 be a measurable subset of S. Then the random countable set*

$$\Pi_1 = \Pi \cap S_1 \tag{2.22}$$

can be regarded either as a Poisson process on S with mean measure

$$\mu_1(A) = \mu(A \cap S_1) \tag{2.23}$$

or as a Poisson process on S_1 whose mean measure is the restriction of μ to S_1.

2.3 The Mapping Theorem

The second great property of Poisson processes is that, if the state space is mapped into another space, the transformed random points again form a Poisson process. The only thing that can go wrong is that the function may pile distinct points on top of one another, and this possibility can be detected by looking at the way the mean measure transforms. Though simple, this property has profound implications.

Thus let Π be a Poisson process on the state space S, having mean measure μ, and let f be a function from S into another (or the same) space T. We assume that T, like S, is a measurable space satisfying the conditions of Section 2.1, and that f is measurable in the sense that

$$f^{-1}(B) = \{x \in S;\ f(x) \in B\}$$

is a measurable subset of S for every measurable $B \subseteq T$.

The points $f(x)$ for $x \in \Pi$ form a random countable set $f(\Pi) \subseteq T$ and we seek to prove that this is a Poisson process in T. To this end, consider the number

$$N^*(B) = \#\{f(\Pi) \cap B\} \tag{2.24}$$

of points of $f(\Pi)$ in B. *So long as the points* $f(x)$ $(x \in \Pi)$ *are distinct,*

$$N^*(B) = \#\{x \in \Pi;\ f(x) \in B\} = N(f^{-1}(B)) \tag{2.25}$$

which has distribution $\mathscr{P}(\mu^*)$, where

$$\mu^* = \mu^*(B) = \mu(f^{-1}(B)). \tag{2.26}$$

Moreover, if $B_1, B_2, \ldots, B_k$ are disjoint, so are their inverse images, so that the $N^*(B_j)$ are independent.

Thus $f(\Pi)$ is a Poisson process in T so long as the points $f(x)$ $(x \in \Pi)$ are distinct. The mean measure is given by (2.26), so that μ^* is the measure induced from μ by the function f. The distinctness condition is however non-trivial, as is obvious if we take f to be the constant function which maps the whole of S, and so the whole of Π, onto a single point of T.

More generally, if μ and f are such that the induced measure has an atom at $t \in T$, then $A = f^{-1}\{t\}$ has non-zero measure $m = \mu(A) = \mu^*(\{t\})$ and there are, with non-zero probability $1 - e^{-m} - m\,e^{-m}$, at least two points of Π which fall in A and are thus both mapped onto t. Thus we need at least to assume that μ^* is non-atomic.

We also need a finiteness condition, and it turns out to be enough for μ to be σ-finite, in the sense that S can be written as a countable union of measurable sets S_n with $\mu(S_n) < \infty$. It is not necessary to assume that μ^* is σ-finite, so that the theorem is strong enough, for example, to cover the process $\phi(\Pi)$ of the last section.

Mapping Theorem *Let Π be a Poisson process with σ-finite mean measure μ on the state space S, and let $f\colon S \to T$ be a measurable function such that the induced measure* (3) *has no atoms. Then $f(\Pi)$ is a Poisson process on T having the induced measure μ^* as its mean measure.*

Proof Suppose first that $\mu(S) < \infty$. Let A be any measurable set and A^c its complement. Let Π_1 and Π_2 denote the restrictions of Π to A and A^c, so that Π_1 and Π_2 are independent Poisson processes. The argument used in the proof of the Disjointness Lemma, now applied to the random set

$$f(\Pi_1) \times f(\Pi_2) \subseteq T \times T$$

shows that $f(\Pi_1)$ and $f(\Pi_2)$ are disjoint with probability 1.

Denote by M the measure on $S \times S$ defined by the recipe that $M(C)$ $(C \subseteq S \times S)$ is the expected number of pairs $(x, y) \in C$ for which

$$x \in \Pi, \qquad y \in \Pi, \qquad f(x) = f(y).$$

Then the disjointness of $f(\Pi_1)$ and $f(\Pi_2)$ shows that

$$M(A \times A^c) = 0, \tag{2.27}$$

for all A. For any $A, B \subseteq S$, $A \times B$ can be written as the union of $(A \cap B) \times (A \cap B)$ and sets lying in either $A \times A^c$ or $B^c \times B$, so that

$$M(A \times B) = m(A \cap B), \tag{2.28}$$

where we have written $m(A) = M(A \times A)$.

Putting $B = S$ in (2.28), $m(A) = M(A \times S)$ which shows that m is a measure. Let M_1 be the measure induced from m on the diagonal $D \subseteq S \times S$ by the function $x \mapsto (x, x)$. Then

$$M_1(A \times B) = m(A \cap B),$$

so that

$$M_1(A \times B) = M(A \times B)$$

for all A, B. By the uniqueness of measure extensions, $M_1 = M$, and in particular M is concentrated on D. Hence, with probability 1, there are no distinct $x \in \Pi$, $y \in \Pi$ with $f(x) = f(y)$. This implies (2.25), from which the result follows.

Now abandon the assumption $\mu(S) < \infty$, and suppose instead that there are disjoint S_n with

$$S = \bigcup_{n=1}^{\infty} S_n, \qquad \mu(S_n) < \infty. \tag{2.29}$$

Let Π_n be the restriction of Π to S_n, so that the Π_n are independent Poisson processes with finite mean measures μ_n. Then the $f(\Pi_n)$ are independent Poisson processes with mean measures $\mu_n^*(B) = \mu_n(f^{-1}(B))$, and the superposition

$$f(\Pi) = f\left(\bigcup_{n=1}^{\infty} \Pi_n\right) = \bigcup_{n=1}^{\infty} f(\Pi_n)$$

is a Poisson process with mean measure

$$\mu^*(B) = \sum_{n=1}^{\infty} \mu_n^*(B) = \mu(f^{-1}(B)).$$

Hence the theorem is proved.

To illustrate the use of the theorem, suppose that Π is a Poisson process in $\mathbb{R}^D$ with rate function $\lambda(x_1, x_2, \ldots, x_D)$.

Take $f\colon \mathbb{R}^D \to \mathbb{R}^d$ to be the projection

$$f(x_1, x_2, \ldots, x_D) = (x_1, x_2, \ldots, x_d) \tag{2.30}$$

where $d \leqslant D$. For $B \subseteq \mathbb{R}^d$,

$$\mu^*(B) = \int_{B \times \mathbb{R}^{D-d}} \cdots \int \lambda(x_1, x_2, \ldots, x_D)\, dx_1\, dx_2 \cdots dx_D$$

$$= \int_B \cdots \int \lambda^*(x_1, x_2, \ldots, x_d)\, dx_1\, dx_2 \cdots dx_d$$

where

$$\lambda^*(x_1, x_2, \ldots, x_d) = \int_{\mathbb{R}^{D-d}} \cdots \int \lambda(x_1, x_2, \ldots, x_D)\, dx_{d+1}\, dx_{d+2} \cdots dx_D. \tag{2.31}$$

Clearly μ is σ-finite and μ^* is non-atomic, so that the theorem applies and we may conclude that, if the integral (2.31) converges, then $f(\Pi)$ is a Poisson process with rate λ^* on $\mathbb{R}^d$. Thus the projection of a higher-dimensional Poisson process is a Poisson process, and the rate of the projection is obtained by integrating out the unwanted variables.

As another example, suppose that Π is a uniform process of rate λ in the plane, and transform from coordinates (x, y) to polar coordinates (x, y). Then f is given by

$$f(x, y) = ((x^2 + y^2)^{1/2}, \tan^{-1}(y/x)),$$

and

$$\mu^*(B) = \iint_{f^{-1}(B)} \lambda\, dx\, dy = \iint_B \lambda r\, dr\, d\theta.$$

Thus the points (r, θ) form a Poisson process in the strip

$$\{(r, \theta);\, r > 0, 0 \leqslant \theta < 2\pi\} \tag{2.32}$$

with rate function

$$\lambda^*(r, \theta) = \lambda r. \tag{2.33}$$

If the angles θ are ignored, the values of r form a Poisson process on $(0, \infty)$ with rate function

$$\lambda^{**}(r) = \int_0^{2\pi} \lambda^*(r, \theta)\, d\theta = 2\pi\lambda r. \tag{2.34}$$

This makes it possible to draw conclusions about the respective distances from the origin of the points of Π. It is for instance easy to show that the distance of the nearest point has a probability density

$$2\pi\lambda r\, e^{-\lambda\pi r^2} \quad (r > 0). \tag{2.35}$$

As a final example, consider the non-homogeneous Poisson process Π on $\mathbb{R}$ governed by the function (2.11), so that M is a continuous increasing function and

$$\mu(a, b] = M(b) - M(a). \tag{2.36}$$

Take f to be the function M itself, which maps $\mathbb{R}$ monotonically into $\mathbb{R}$. If $t > 0$ and $x = M(t)$, then

$$\mu^*(0, x] = \mu^*(0, M(t)] = \mu(0, t] = M(t) = x,$$

and the same argument shows that $\mu^*(-x, 0] = x$. Thus $M(\Pi)$ is a uniform Poisson process of rate 1 on the (finite or infinite) interval $(M(-\infty), M(\infty))$.

It follows that every one-dimensional Poisson process can be transformed into one of constant rate, by means of a continuous monotonic transformation. This means that in one dimension only the uniform Poisson process is of fundamental importance (see Chapter 4), and the properties of the most general locally finite process can be inferred from it.

2.4 The Bernoulli process

Let Π be a Poisson process on S with mean measure μ, and suppose that $\mu(S) < \infty$. Then Π is almost certainly a finite subset of S, the total number $N(S)$ having the Poisson distribution $\mathscr{P}(\mu(S))$. What happens if we condition Π on the value of $N(S)$?

Denote by $\mathbb{P}_n$ the conditional probability measure

$$\mathbb{P}_n\{\cdot\} = \mathbb{P}\{\cdot|N(S) = n\}. \tag{2.37}$$

Let $A_1, A_2, \ldots, A_k$ be disjoint subsets of S. Then

$$\begin{aligned}
\mathbb{P}_n\{N(A_1) = n_1, N(A_2) = n_2, \ldots, N(A_k) = n_k\} \\
&= \mathbb{P}\{N(A_0) = n_0, N(A_1) = n_1, \ldots, N(A_k) = n_k\}/\mathbb{P}\{N(S) = n\} \\
&= \prod_{j=0}^{k} \frac{e^{-\mu(A_j)}\mu(A_j)^{n_j}}{n_j!} \Bigg/ \frac{e^{-\mu(S)}\,\mu(S)^n}{n!} \\
&= \frac{n!}{n_0!\, n_1! \cdots n_k!} \left(\frac{\mu(A_0)}{\mu(S)}\right)^{n_0} \left(\frac{\mu(A_1)}{\mu(S)}\right)^{n_1} \cdots \left(\frac{\mu(A_k)}{\mu(S)}\right)^{n_k}
\end{aligned}$$

where $n_0 = n - \sum_1^k n_j$ and A_0 is the complement of $A_1 \cup A_2 \cup \cdots \cup A_k$.

A random finite set $\Pi \subseteq S$ for which $\#\Pi = n$ and

$$\mathbb{P}\{N(A_1) = n_1, N(A_2) = n_2, \ldots, N(A_k) = n_k\}$$
$$= \frac{n!}{n_0!\, n_1! \cdots n_k!} p(A_0)^{n_0} p(A_1)^{n_1} \cdots p(A_k)^{n_k} \quad (2.38)$$

is called a *Bernoulli process*. Its mean measure is

$$\mathbb{E}\{N(A)\} = np(A), \quad (2.39)$$

so that p is a probability measure on S. With this terminology we can say that conditioning on $N(S) = n$ converts the Poisson process with finite mean measure μ into the Bernoulli process with parameters n and

$$p(\cdot) = \mu(\cdot)/\mu(S). \quad (2.40)$$

There is a much easier way of constructing a Bernoulli process. Let $X_1, X_2, \ldots, X_n$ be independent random variables, distributed over the space S according to the probability measure p. If p has no atoms the X_r are distinct with probability one (the formal proof is a simpler version of that of the Disjointness Lemma), so that $\{X_1, X_2, \ldots, X_n\}$ is a random set with n elements. It is then an easy multinomial calculation that the counts

$$N(A) = \#\{r;\, X_r \in A\} \quad (2.41)$$

satisfy (2.38).

Thus, given $N(S)$, the points of a finite Poisson process look exactly like $N(S)$ independent random variables, with common distribution (2.40). This fact has a number of consequences, the most important of which is described in the next section. Here is another: consider a random variable $Z = Z(\Pi)$ defined in terms of the Poisson process Π, and suppose that we can calculate its expectation $\mathbb{E}(Z)$ in terms of the mean measure μ, assumed finite. Set $\mu(\cdot) = \mu p(\cdot)$ where p is a fixed probability measure and $\mu = \mu(S)$ is allowed to vary. Then $\mathbb{E}(Z)$ is a function of μ: $\mathbb{E}(Z) = \psi(\mu)$, say. We can then write

$$\mathbb{E}(Z) = \sum_{n=0}^{\infty} \pi_n(\mu)\mathbb{E}_n(Z), \quad (2.42)$$

where

$$\mathbb{E}_n(Z) = \mathbb{E}(Z \mid N(S) = n)$$

is the corresponding expectation for the Bernoulli process (2.38) and *does not depend on* μ. Hence

$$\psi(\mu)\,\mathrm{e}^{\mu} = \sum_{n=0}^{\infty} \frac{\mu^n}{n!} \mathbb{E}_n(Z). \quad (2.43)$$

By expanding $\psi(\mu)\,e^{\mu}$ as a power series in μ we can read off the $\mathbb{E}_n(Z)$ from the successive coefficients. An example is worked in detail in Section 4.3.

2.5 The Existence Theorem

We can now reverse the argument of the last section to prove the existence of Poisson processes with given mean measures. We need of course to assume that the given measure is non-atomic, and we also require a weak finiteness condition. Here σ-finiteness is sufficient but not necessary, and it is worth going a little wider to include measures like (2.18).

Existence Theorem *Let μ be a non-atomic measure on S which can be expressed in the form*

$$\mu = \sum_{n=1}^{\infty} \mu_n, \qquad \mu_n(S) < \infty. \tag{2.44}$$

Then there exists a Poisson process on S having μ as its mean measure.

Proof Without loss of generality, suppose $\mu_n(S) > 0$ for all n. On a suitable probability space construct independent random variables

$$N_n, \qquad X_{nr} \quad (n, r = 1, 2, 3, \ldots,)$$

such that the distribution of N_n is $\mathscr{P}(\mu_n(S))$ and the distribution of X_{nr} is

$$p_n(\cdot) = \mu_n(\cdot)/\mu_n(S). \tag{2.45}$$

Write

$$\Pi_n = \{X_{n1}, X_{n2}, \ldots, X_{nN_n}\} \tag{2.46}$$

and

$$\Pi = \bigcup_{n=1}^{\infty} \Pi_n. \tag{2.47}$$

Writing

$$N_n(A) = \#\{\Pi_n \cap A\}$$

we have, for disjoint $A_1, A_2, \ldots, A_k$ and writing A_0 for the complement of their union,

$$\mathbb{P}\{N_n(A_1) = m_1, N_n(A_2) = m_2, \ldots, N_n(A_k) = m_k \mid N_n = m\}$$
$$= \frac{m!}{m_0!\, m_1! \cdots m_k!}\, p_n(A_0)^{m_0} p_n(A_1)^{m_1} \cdots p_n(A_k)^{m_k}$$

if $m_0 = m - m_1 - m_2 - \cdots - m_k$. Hence

$$\mathbb{P}\{N_n(A_1) = m_1, N_n(A_2) = m_2, \ldots, N_n(A_k) = m_k\}$$

$$= \sum_{m=\sum m_j}^{\infty} \frac{e^{-\mu_n(S)}\mu_n(S)^m}{m!} \frac{m!}{m_0!\, m_1! \cdots m_k!} \prod_{j=0}^{k} p_n(A_j)^{m_j}$$

$$= \sum_{m=\sum m_j}^{\infty} \pi_{m-\sum m_j}(\mu_n(A_0)) \prod_{j=0}^{k} \pi_{m_j}(\mu_n(A_j))$$

$$= \prod_{j=0}^{k} \pi_{m_j}(\mu_n(A_j)).$$

Thus the $N_n(A_j)$ are independent random variables with distributions $\mathscr{P}(\mu_n(A_j))$, so that the Π_n are independent Poisson processes with respective mean measures μ_n. The Superposition Theorem shows that (2.47) defines a Poisson process with mean measure μ, and the proof is complete.

3

Sums over Poisson processes

3.1 Means, variances, and distributions

This chapter is devoted to sums of the form

$$\Sigma = \sum_{X \in \Pi} f(X) \tag{3.1}$$

where f is a real-valued function on the state space of a Poisson process Π. There are a number of simple, but very important, results which give conditions for the (absolute) convergence of such sums, and expressions for the expectation, variance, and characteristic function of the random variable Σ. In this section we shall derive the form of these results without rigorous proof, the latter being deferred to the next section.

Sums of the form (3.1) occur in many applications of Poisson processes. One of the earliest was the so-called *shot effect*, in which the points of a Poisson process in one (time) dimension have an effect that continues for a time after the event represented by each random point. Suppose for example that particles are emitted from a lump of radioactive material at instants $0 < T_1 < T_2 < \cdots$ which form a Poisson process on $(0, \infty)$. Suppose that these are recorded in a device which responds at time t to a particle at T_j by a measurable quantity of the form $\phi(t - T_j)$ for some function ϕ. If these quantities are added in the device, the response will be

$$\sum_j \phi(t - T_j),$$

which for fixed t is of the form (3.1).

For an example in several dimensions, suppose that stars in some region of space are scattered according to a Poisson process π and that we want to calculate the resulting gravitational field. The gravitational potential at a point x is then of the form (3.1), where $f(x) = |x - X|^{-1}$ so long as the stars are of the same mass (cf. Section 5.3).

The technique for finding the distribution of Σ is first to calculate it for simple functions f and then to extend the result to more complicated f by the standard techniques of integration theory.

Consider first a function $f\colon S \to \mathbb{R}$ which takes only a finite number of non-zero values $f_1, f_2, \ldots, f_k$ and is zero outside a set for which the mean measure μ of Π is finite. Then

$$A_j = \{x; f(x) = f_j\} \tag{3.2}$$

is measurable with

$$m_j = \mu(A_j) < \infty, \tag{3.3}$$

and the A_j are disjoint. Thus

$$N_j = N(A_j) \tag{3.4}$$

are independent with respective distributions $\mathscr{P}(m_j)$, and

$$\Sigma = \sum_{X \in \Pi} f(X) = \sum_{j=1}^{k} f_j N_j. \tag{3.5}$$

This determines the distribution of Σ, most conveniently through its characteristic (or moment generating) function. For any real or complex θ,

$$\begin{aligned}\mathbb{E}(e^{\theta\Sigma}) &= \prod_{j=1}^{k} \mathbb{E}(e^{\theta f_j N_j}) \\ &= \prod_{j=1}^{k} \exp\{(e^{\theta f_j} - 1)m_j\} \\ &= \exp\left\{\sum_{j=1}^{k} \int_{A_j} (e^{\theta f(x)} - 1)\mu(dx)\right\} \\ &= \exp\left\{\int_S (e^{\theta f(x)} - 1)\mu(dx)\right\}\end{aligned}$$

since $f = 0$ outside the union of the A_j. This equation

$$\mathbb{E}(e^{\theta\Sigma}) = \exp\left\{\int_S (e^{\theta f(x)} - 1)\mu(dx)\right\} \tag{3.6}$$

is the master equation from which all else follows.

If f can take positive and negative values, it is best to take θ pure imaginary, so that (3.6) becomes

$$\mathbb{E}(e^{it\Sigma}) = \exp\left\{\int_S (e^{itf(x)} - 1)\mu(dx)\right\} \tag{3.7}$$

for real t. The left-hand side is the characteristic function of Σ, a knowledge of which for all real t determines the distribution of Σ. It will be shown in the next section that, for any measurable f, the sum (3.1) is absolutely convergent with probability 1 if and only if the integral in (3.7) converges absolutely (for some and then for all real $t \neq 0$), and that (3.7) then holds.

If f takes only positive values, it is more natural to take θ real and negative, so that (3.6) becomes

$$\mathbb{E}(e^{-u\Sigma}) = \exp\left\{-\int_S (1 - e^{-uf(x)})\mu(dx)\right\} \tag{3.8}$$

for all $u \geqslant 0$. In this form it is true without restriction; if the integral diverges $\Sigma = \infty$ with probability 1, and (3.8) takes the trivial form $0 = 0$.

Expanding (3.6) as a power series in θ and equating coefficients of θ and θ^2 we obtain the simple identities

$$\mathbb{E}(\Sigma) = \int_S f(x)\mu(dx) \tag{3.9}$$

and

$$\operatorname{var}(\Sigma) = \int_S f(x)^2\mu(dx). \tag{3.10}$$

If f_1 and f_2 are two real-valued functions, replace θf by $\theta_1 f_1 + \theta_2 f_2$ in (3.6), to obtain the equation

$$\mathbb{E}(e^{\theta_1\Sigma_1 + \theta_2\Sigma_2}) = \exp\left\{\int_S (e^{\theta_1 f_1(x) + \theta_2 f_2(x)} - 1)\mu(dx)\right\}, \tag{3.11}$$

which determines the joint distribution of

$$\Sigma_1 = \sum_{X\in\Pi} f_1(X), \qquad \Sigma_2 = \sum_{X\in\Pi} f_2(X).$$

The result extends at once in an obvious way to three or more functions.

The coefficient of $\theta_1\theta_2$ in the power series expansion of (3.11) gives

$$\mathbb{E}(\Sigma_1\Sigma_2) = \int f_1\, d\mu \int f_2\, d\mu + \int f_1 f_2\, d\mu, \tag{3.12}$$

the second term on the right representing therefore the covariance of Σ_1 and Σ_2. It is sometimes useful to write the left-hand side as

$$\mathbb{E}\left(\sum_{X_1,X_2\in\Pi} f_1(X_1)f_2(X_2)\right) = \mathbb{E}\left(\sum_{X\in\Pi} f_1(X)f_2(X)\right) + \mathbb{E}\left(\sum_{X_1\neq X_2} f_1(X_1)f_2(X_2)\right).$$

The first term on the right is $\int f_1 f_2\, d\mu$, so (3.12) shows that

$$\mathbb{E}\left(\sum_{\substack{X_1,X_2\in\Pi\\ X_1\neq X_2}} f_1(X_1)f_2(X_2)\right) = \mathbb{E}\left(\sum_{X_1\in\Pi} f_1(X_1)\right)\mathbb{E}\left(\sum_{X_2\in\Pi} f_2(X_2)\right). \tag{3.13}$$

This identity generalises to products of any number of functions:

$$\mathbb{E}\left(\sum_{\substack{X_1,X_2,\ldots,X_n\in\Pi\\ X_j \text{ distinct}}} f_1(X_1)f_2(X_2)\cdots f_n(X_n)\right) = \prod_{j=1}^{n} \mathbb{E}\left(\sum_{X_j\in\Pi} f_j(X_j)\right). \tag{3.14}$$

All these results stem from pioneering work of N. R. Campbell (1909, 1910), and particular cases are known in the literature as *Campbell's Theorem.* We shall prove them all, under this name, in the next section. They provide a simple means of evaluating a variety of distributions associated with Poisson processes. In most applications, μ will have a density λ with respect to Lebesgue measure on S, so that integrals with respect to μ are to be read

$$\int_S \cdots \mu(\mathrm{d}x) = \int_S \cdots \lambda(x)\,\mathrm{d}x$$

as in Section 2.1.

The results are even more general than they look. It will be seen in Chapter 5 that they generalise at once to situations like the stellar example in which the star masses are not equal but are independent random variables.

3.2 Campbell's Theorem

The formal calculations of the last section are justified by taking the identity (3.6) for simple functions f and extending it to more general functions by integration theory. The results are drawn together in the following theorem.

Campbell's Theorem *Let Π be a Poisson process on S with mean measure μ, and let $f\colon S \to \mathbb{R}$ be measurable. Then the sum*

$$\Sigma = \sum_{X \in \Pi} f(X) \tag{3.15}$$

is absolutely convergent with probability if and only if

$$\int_S \min(|f(x)|, 1)\,\mu(\mathrm{d}x) < \infty. \tag{3.16}$$

If this condition holds, then

$$\mathbb{E}(\mathrm{e}^{\theta\Sigma}) = \exp\left\{\int_S (\mathrm{e}^{\theta f(x)} - 1)\,\mu(\mathrm{d}x)\right\} \tag{3.17}$$

for any complex θ for which the integral on the right converges, and in particular whenever θ is pure imaginary. Moreover

$$\mathbb{E}(\Sigma) = \int_S f(x)\,\mu(\mathrm{d}x) \tag{3.18}$$

in the sense that the expectation exists if and only if the integral converges, and they are then equal. If (3.18) *converges, then*

$$\operatorname{var}(\Sigma) = \int_S f(x)^2 \mu(\mathrm{d}x), \tag{3.19}$$

finite or infinite.

Proof We already know that (3.17) holds, for all complex θ, if f is a function taking a finite number of values and vanishing outside a set of finite μ-measure. Call such functions *simple.*

Any positive measurable function f may be expressed as the limit of an increasing sequence (f_j) of simple functions. Thus taking $\theta = -u$ real and negative, and setting $\Sigma_j = \sum f_j(X)$ we have

$$\begin{aligned}\mathbb{E}(\mathrm{e}^{-u\Sigma}) &= \lim_{j\to\infty} \mathbb{E}(\mathrm{e}^{-\theta\Sigma_j})\\ &= \lim_{j\to\infty} \exp\left\{-\int (1 - \mathrm{e}^{-uf_j(x)})\mu(\mathrm{d}x)\right\}\\ &= \exp\left\{-\int (1 - \mathrm{e}^{-uf(x)})\mu(\mathrm{d}x)\right\}\end{aligned}$$

for $u > 0$ by monotone convergence. If (3.16) holds, the last integral converges and tends to 0 as $u \to 0$, showing that Σ is a finite random variable. Then both sides of (3.17) are analytic functions of θ in $\operatorname{Re}\theta \leqslant 0$, so that (3.17) holds for $\operatorname{Re}\theta \leqslant 0$.

On the other hand, if (3.16) does not hold, the last integral diverges for all $u > 0$, so that $\mathbb{E}(e^{-u\Sigma}) = 0$, showing that $\Sigma = \infty$ with probability 1. An exactly similar argument establishes (3.18) and (3.19).

Thus the theorem is proved if $f \geqslant 0$. To complete the proof, apply the theorem to the positive functions

$$f^+ = \max(f, 0), \qquad f^- = \max(-f, 0).$$

The sum converges absolutely if and only if

$$\Sigma_+ = \sum_{X\in\Pi} f^+(X) = \sum_{X\in\Pi_+} f(X)$$

and

$$\Sigma_- = \sum_{X\in\Pi} f^-(X) = -\sum_{X\in\Pi_-} f(X)$$

converge, where Π_+ and Π_- are the restrictions of Π to

$$S_+ = \{f > 0\}, \qquad S_- = \{f < 0\}.$$

This shows that (3.16) is the necessary and sufficient condition for convergence. If it holds, and θ is pure imaginary, then

$$\begin{aligned}\mathbb{E}(e^{\theta\Sigma}) &= \mathbb{E}(e^{\theta\Sigma_+ - \theta\Sigma_-})\\ &= \mathbb{E}(e^{\theta\Sigma_+})\mathbb{E}(e^{-\theta\Sigma_-})\\ &= \exp\left\{\int (e^{\theta f^+} - 1)\,d\mu\right\}\exp\left\{\int (e^{-\theta f^-} - 1)\,d\mu\right\}\\ &= \exp\left\{\int (e^{\theta f^+} + e^{-\theta f^-} - 2)\,d\mu\right\}\\ &= \exp\left\{\int (e^{\theta f} - 1)\,d\mu\right\},\end{aligned}$$

where we have used the fact that Π_+ and Π_- are independent Poisson processes, being the restrictions of Π to disjoint sets.

Thus (3.17) holds at least for imaginary θ and its extension to all complex θ for which the integral converges is a matter of analytic continuation. Again the proof of (3.18) and (3.19) is similar.

Corollary 3.1 *If $f_1, f_2, \ldots, f_n$ are functions satisfying* (3.16), *so that*

$$\Sigma_j = \sum_{X \in \Pi} f_j(X) \tag{3.20}$$

converges with probability 1, *then*

$$\mathbb{E}(e^{it_1\Sigma_1 + it_2\Sigma_2 + \cdots + it_n\Sigma_n}) = \exp\left(\int_S (e^{it_1 f_1(x) + \cdots + it_n f_n(x)} - 1)\mu(dx)\right\}. \tag{3.21}$$

If the f_j satisfy

$$\int_S f_j(x)^2 \mu(dx) < \infty, \tag{3.22}$$

then

$$\operatorname{cov}(\Sigma_j, \Sigma_k) = \int_S f_j(x) f_k(x) \mu(dx). \tag{3.23}$$

Proof Replace f by $t_1 f_1 + \cdots + t_n f_n$ in (3.17) with $\theta = i$, (3.18) and (3.19).

The convergence condition (3.16) can usefully be expressed in terms of the function

$$\gamma(a) = \mu\{x \in S; |f(x)| \geqslant a\}. \tag{3.24}$$

It is easy to check that (3.16) is equivalent to the condition that $\gamma(a)$ be finite for all $a > 0$, and that

$$\int_0^1 \gamma(a)\,da < \infty. \tag{3.25}$$

Finally, it is worth noting that (3.19) may be finite even if (3.18) does not converge. Indeed, it is easy to construct examples in which even (3.16) is violated, so that Σ does not converge absolutely, but (3.19) has a finite integral on the right-hand side. This suggests that some generalised meaning can be attached to the sum (3.15), yielding a random variable with finite variance.

One way of doing this is as follows. Suppose that $\int f^2 \, \mathrm{d}\mu < \infty$ and that S is the union of an increasing sequence of subsets S_n such that (for instance by symmetry)

$$\int_{S_n} f(x)\mu(\mathrm{d}x) = 0. \tag{3.26}$$

Write

$$\Sigma_n = \sum_{X \in \Pi \cap S_n} f(X), \tag{3.27}$$

so that Σ_n is a finite random variable with zero mean and finite variance. It is then simple to show that $\mathbb{E}\{(\Sigma_m - \Sigma_n)^2\} \to 0$ as $m, m \to \infty$. It follows that there is a random variable Σ which is the mean square limit of Σ_n:

$$\mathbb{E}\{(\Sigma_n - \Sigma)^2\} \to 0 \tag{3.28}$$

as $n \to \infty$. This random variable has zero mean and finite variance given by (3.19).

We shall meet a physical application of this argument in Section 5.3.

3.3 The characteristic functional

Specialising Campbell's Theorem to the case $f \geqslant 0$ and setting $\theta = -1$, (3.17) becomes

$$\mathbb{E}(\mathrm{e}^{-\Sigma_f}) = \exp\left\{-\int_S (1 - \mathrm{e}^{-f(x)})\mu(\mathrm{d}x)\right\}, \tag{3.29}$$

where

$$\Sigma_f = \sum_{X \in \Pi} f(X). \tag{3.30}$$

The quantity on the left of (3.29), regarded as a function of the arbitrary function f is an example of what is called the *characteristic functional* of a random process.

The name implies that, in a certain sense, equation (3.29) characterises the Poisson process. Suppose that Π is *any* random countable subset of S, and define $\Sigma_f \leqslant \infty$ for any function $f\colon S \to [0, \infty)$ by (3.30). Suppose that, for some measure μ, (3.29) holds for a certain class $\mathscr{F}$ of functions. If $\mathscr{F}$ is rich enough to contain all functions that take on only a finite number of different values $f_1, f_2, \ldots, f_k$ then (3.29) gives

$$\mathbb{E}(e^{-\Sigma_f}) = \exp\left\{-\sum_{j=1}^{k} (1 - e^{-f_j})m_j\right\}, \tag{3.31}$$

with

$$m_j = \mu(A_j), \qquad A_j = \{x_j; f(x) = f_j\}$$

and

$$\Sigma_f = \sum_{j=1}^{k} f_j N(A_j).$$

Thus, with $z_j = e^{-f_j}$,

$$\mathbb{E}(z_1^{N(A_1)} z_2^{N(A_2)} \ldots z_k^{N(A_k)}) = \prod_{j=1}^{k} e^{m_j(z-1)}. \tag{3.32}$$

Since this holds for $0 < z_j < 1$, the $N(A_j)$ are independent, with Poisson distributions $\mathscr{P}(m_j)$ (and the obvious conventions if $m_j = \infty$). Thus Π is a Poisson process with mean measure μ.

In order to prove that a random countable set is a Poisson process, it therefore suffices to establish (3.29) for a sufficiently rich class of functions.

A remarkable characterisation of the homogeneous Poisson process in one dimension was proved by Rényi in 1967. He showed that, if a random subset of $\mathbb{R}$ has the property that, for any set A which is a finite union of intervals, the count $N(A)$ has a Poisson distribution with mean $\lambda|\mathrm{A}|$, then that random set is a Poisson process with constant rate λ.

This theorem is surprising because it makes no assumption of independence of the counts $N(A)$ for disjoint A_j; that emerges as a consequence of the Poisson assumption. On the other hand, a counterexample due to Moran (1967) shows that it is essential to require the Poisson distribution for finite unions. His example is a non-Poisson process such that $N(A)$ has distribution $\mathscr{P}(\lambda|A|)$ for every interval A.

One can see that Rényi's result is plausible from the characteristic functional approach, even though this does not seem to yield a complete proof. To prove that a random subset of $\mathbb{R}$ is a Poisson process of constant rate λ, it suffices to prove the special case of (3.29), that

$$\mathbb{E}(e^{-\Sigma_f}) = \exp\left\{-\int_{-\infty}^{\infty} (1 - e^{-f(x)})\lambda \, dx\right\}, \tag{3.33}$$

for all positive step functions f, that is, for all functions with a finite number of discontinuities, constant between neighbouring discontinuities, and zero outside a finite set.

The assumption that $N(A)$ has the Poisson distribution $\mathscr{P}(\lambda|A|)$ for all finite unions of intervals is equivalent to supposing that (3.33) holds for all step functions taking only two values. Hence the force of Rényi's result is that we can extend (3.33) from two-valued step functions to all step functions.

At first sight this seems unlikely, but it becomes less so if we note that, from the point of view of integration theory, two-valued step functions can approximate general step functions. Let f be a positive step function, and consider an interval (a, b) on which f is constant:

$$f(x) = f \quad (a < x < b).$$

For any constant $M > f$, divide (a, b) into subintervals $I_1, I_2, \ldots, I_{2n}$, where the odd-numbered intervals are of length $(b-a)(M-f)/Mn$ and the even-numbered of length $(b-a)f/Mn$. Let g_n be the two-valued step function on (a, b) equal to 0 on the odd intervals and to M on the even ones. Then it is elementary to check that, for any continuous ϕ,

$$\lim_{n\to\infty} \int_a^b \phi(x) g_n(x)\,\mathrm{d}x = \int_a^b \phi(x) f(x)\,\mathrm{d}x.$$

If M is taken greater than the overall maximum of the step functions f, the g_n on different intervals can be pasted together in an obvious way to produce a two-valued g_n with

$$\lim_{n\to\infty} \int_{-\infty}^{\infty} \phi(x) g_n(x)\,\mathrm{d}x = \int_{-\infty}^{\infty} \phi(x) f(x)\,\mathrm{d}x \tag{3.34}$$

for every continuous ϕ. In this weak sense the two-valued step functions form a dense subset of the set of all step functions.

By imposing suitable conditions on the random set, we can ensure that the left hand side of (3.33) is continuous in f in the sense (3.34), but a general proof is elusive. In the next section Rényi's Theorem will be proved by a quite different argument and in a much more general setting, but the reader may catch faint echoes of this approximation technique.

Before leaving (3.29) it is worth mentioning a slight reformulation that is sometimes useful. Writing $F(x) = \mathrm{e}^{-f(x)}$, (3.29) becomes

$$\mathbb{E}\{\prod F(X)\} = \exp\left\{-\int_S [1 - F(x)]\mu(\mathrm{d}x)\right\}, \tag{3.35}$$

where the product on the left is taken over all points X on the random set, and is to be interpreted in the usual infinite product sense. The translation from f to F establishes (3.35) for functions F with $0 < F \leqslant 1$. However, the argument of the last section can be made to show directly that (3.37) is true without this restriction if the integral on the right is absolutely convergent. It is sometimes helpful to use it when $F = 0$ or takes negative values.

3.4 Rényi's Theorem

Rényi's original result is a characterisation of the homogeneous Poisson process in one dimension, but it can be generalised in several ways. First of all, the mean measure need not be a multiple of Lebesgue measure, and the state space may be more complicated than $\mathbb{R}$. We shall not strive for the most general result, and shall be content to prove the result for a random subset of $\mathbb{R}^d$ which is finite on bounded sets.

A more interesting line of development is suggested by the work of Kendall (1974). In the course of developing a very general theory of random sets, he shows that a great deal of information is contained in the *avoidance function*

$$\alpha(A) = \mathbb{P}\{N(A) = 0\} = \mathbb{P}\{\Pi \cap A = \varnothing\} \tag{3.36}$$

For a Poisson process with mean measure μ,

$$\alpha(A) = e^{-\mu(A)}. \tag{3.37}$$

It turns out not to be necessary to assume that $N(A)$ has a Poisson distribution, but only that it has the right zero-probability (3.37) for some measure μ.

Thus we have the following general version of Rényi's Theorem. Its formulation is in terms of *rectangles* in $\mathbb{R}^d$, by which we mean subsets of the form

$$(a, b] = \{x = (x_1, \ldots, x_d);\, a_i < x_i \leqslant b_i \quad (i = 1, 2, \ldots, d)\}, \tag{3.38}$$

where $a, b \in \mathbb{R}^d$ satisfy $a_i < b_i$ for each i.

Rényi's Theorem *Let μ be a non-atomic measure on $\mathbb{R}^d$, finite on bounded sets. Let Π be a random countable subset of $\mathbb{R}^d$ such that, whenever A is a finite union of rectangles,*

$$\mathbb{P}\{\Pi \cap A = 0\} = e^{-\mu(A)}. \tag{3.39}$$

Then Π is a Poisson process with mean measure μ.

Before proving the theorem, we note that there are two distinct ways of applying it. This duality shows the subtle interplay between the Poisson distribution and independence.

The first starts from a random countable subset Π with the property that, for any finite union A of rectangles, the count $N(A)$ has a Poisson distribution with finite mean. No assumption is made of independence. Then

$$\mu(A) = \mathbb{E}\{N(A)\} \tag{3.40}$$

is, by the arguments of Section 2.1, a non-atomic measure on $\mathbb{R}^d$ which is finite on bounded sets. Because $N(A)$ has distribution $\mathscr{P}(\mu(A))$, (3.39) holds, and the theorem shows that Π is a Poisson process.

One can however proceed quite differently, with no mention at all of the Poisson distribution. Suppose that Π is a random countable set, and that the avoidance function satisfies

$$\alpha(A) > 0 \qquad \text{if } A \text{ is bounded} \tag{3.41}$$

Suppose also that, if A and B are finite unions of rectangles and are disjoint, then the events

$$\{\Pi \cap A = \varnothing\}, \qquad \{\Pi \cap B = \varnothing\}$$

are independent. Then

$$\alpha(A \cup B) = \mathbb{P}\{\Pi \cap A = \Pi \cap B = \varnothing\} = \alpha(A)\alpha(B), \tag{3.42}$$

so that

$$\mu(A) = -\log \alpha(A) \tag{3.43}$$

is finitely additive. If we can make the step to show that μ is countably additive and non-atomic, the theorem applies, and Π is a Poisson process. (This strengthens the remarks in Section 1.4 about the inevitability of the Poisson process.)

It is straightforward to show that the finitely additive μ is a non-atomic measure if, whenever (A_n) is a decreasing sequence of rectangles the length of whose sides tends to zero (so that their intersection is either empty or a single point), $\lim \mu(A_n) = 0$. Hence the only further condition we need to impose is that, for any such sequence,

$$\lim \alpha(A_n) = 1. \tag{3.44}$$

Such a condition is necessary to exclude sets Π having fixed points, i.e. points x with

$$\mathbb{P}\{x \in \Pi\} > 0. \tag{3.45}$$

To take the most trivial example, if Π is the random set which is either $\varnothing$ (with probability $\frac{1}{2}$) or $\{0\}$ (with probability $\frac{1}{2}$), then

$$\begin{aligned} \alpha(A) &= 1 \quad (0 \notin A) \\ &= \tfrac{1}{2} \quad (0 \in A) \end{aligned}$$

satisfies (3.41) and (3.42) but not (3.44), and Π is certainly not a Poisson process.

Proof of Rényi's Theorem For the purposes of this proof we describe as a *k-cube* a rectangle $(a, b]$ for which

$$a_i = v_i 2^{-k}, \qquad b_i = (v_i + 1)2^{-k} \quad (1 \leqslant i \leqslant d) \tag{3.46}$$

the v_i being integers. For any k the k-cubes form a dissection of $\mathbb{R}^d$. For any k-cube C write $E(C) = \{\Pi \cap C = \varnothing\}$ for the event that there are no points of the random set Π in the k-cube C. Fix k and let $C_1, C_2, \ldots, C_n$ be k-cubes which are distinct and thus disjoint. Then

$$\mathbb{P}\left\{\bigcap_{r=1}^{n} E(C_r) = \mathbb{P}\{\Pi \cap C_r = \varnothing \quad (r = 1, 2, \ldots, n)\right\}$$
$$= \mathbb{P}\{\Pi \cap (\bigcup C_r) = \varnothing\}$$
$$= \alpha(\bigcup C_r) = \exp\{-\mu(\bigcup C_r)\}$$
$$= \prod_{r=1}^{n} \mathrm{e}^{-\mu(C_r)}.$$

Hence the events $E(C)$ for k-cubes C, are independent with

$$\mathbb{P}\{E(C)\} = \mathrm{e}^{-\mu(C)}. \tag{3.47}$$

We use this fact to compute the distribution of

$$N(G) = \#\{\Pi \cap G\} \tag{3.48}$$

for any bounded open set G. Every point of G lies, for large enough k, in some k-cube contained in G. Moreover, two distinct points of G lie for large enough k in different k-cubes. Hence

$$N(G) = \lim_{k\to\infty} N_k(G), \tag{3.49}$$

where $N_k(G)$ is the number of k-cubes $C \subset G$ for which $E(C)$ does not occur. Note that $N_k(G)$ increases with k.

Because the $E(C)$ are independent, the probability generating function of $N_k(G)$ is

$$\mathbb{E}(z^{N_k(G)}) = \prod \{\mathbb{P}[E(C)] + (1 - \mathbb{P}[E(C)])z\}$$
$$= \prod \{\mathrm{e}^{-\mu(C)} + (1 - \mathrm{e}^{-\mu(C)})z\}$$

in $|z| < 1$, where the product is over all k-cubes $C \subset G$. In particular, taking $0 \leqslant z < 1$, (3.49) and monotone convergence show that

$$\mathbb{E}(z^{N(G)}) = \lim_{k\to\infty} \prod \{z + (1 - z)\mathrm{e}^{-\mu(C)}\} \tag{3.50}$$

with the convention that $z^\infty = 0$.

For such z, and $\mu \geqslant 0$,

$$z + (1 - z)\,\mathrm{e}^{-\mu} \geqslant \mathrm{e}^{-(1-z)\mu}.$$

Hence

$$\mathbb{E}(z^{N(G)}) \geqslant \lim_{k\to\infty} \prod \mathrm{e}^{-(1-z)\mu(C)}$$
$$\geqslant \mathrm{e}^{-(1-z)\mu(G)}, \tag{3.51}$$

since the k-cubes $C \subset G$ are disjoint. Letting $z \to 1$ shows that $N(G)$ is finite with probability 1.

To evaluate the limit in (3.50), we need to use the fact that, for k-cubes in a bounded region, $\mu(C)$ tends to zero uniformly as $k \to \infty$. This is a moderately familiar consequence of the lack of atoms of μ and is best proved by contradiction. It is sufficient to prove it for subsets of a fixed 0-cube C_0. If it is false, there exist $\delta > 0$ and k-cubes $C \subset C_0$ for every value of k with $\mu(C) > \delta$. Call a k-cube *bad* if, for every $l > k$, C contains an l-cube C' with $\mu(C') > \delta$. By hypothesis C_0 is bad. It follows that, of the 2^d 1-cubes in C_0, at least one (C_1, say) is bad. Proceeding thus, we have a sequence $C_0 \supset C_1 \supset C_2 \supset \cdots$ where C_k is a bad k-cube. A bad k-cube contains C' with $\mu(C') > \delta$, and so has $\mu(C) > \delta$. Hence $\mu(C_k) > \delta$ for all k. But since C_k is a k-cube with side 2^{-k}, the intersection of the C_k is either empty or a single point, and has measure at least δ. This is the required contradiction.

Hence, returning to G, we can use the fact that $\mu(C) \leqslant \delta$ for all k-cubes in G with sufficiently large k. If $\mu \leqslant \delta$ and z is fixed

$$e^{-(1-z)\mu} \leqslant z + (1-z)\,e^{-\mu} \leqslant e^{-(1-z)\psi(\delta)\mu}$$

where $\psi(\delta) \to 1$ as $\delta \to 0$. Hence

$$e^{-(1-z)\mu(G)} \leqslant \mathbb{E}(z^{N(G)}) \leqslant e^{-(1-z)\psi(\delta)\mu(G)},$$

and letting $\delta \to 0$,

$$\mathbb{E}(z^{N(G)}) = e^{-(1-z)\mu(G)}. \tag{3.52}$$

This holds for all z in $0 \leqslant z < 1$, and so $N(G)$ has a Poisson distribution with mean $\mu(G)$. Now let $G_1, G_2, \ldots, G_m$ be disjoint bounded open sets. No k-cube can be contained in more than one G_j and hence the variables

$$N_k(G_1),\ N_k(G_2), \ldots, N_k(G_m)$$

are independent for fixed k. Letting $k \to \infty$ shows that the variables

$$N(G_1),\ N(G_2), \ldots, N(G_m)$$

are independent.

To complete the proof, we need to extend these results from bounded open sets to general Borel sets. First notice that they are true for rectangles (3.38), since they are intersections of sequences of open sets. It follows that (3.29) holds whenever f is a function which is a linear combination of the functions χ_r, where each χ_r is 1 on a particular rectangle, and zero outside. Any positive measurable function is the limit of an increasing sequence of such combinations, and therefore satisfies (3.29). This is enough to prove that Π is a Poisson process, with mean measure μ, as asserted.

4

Poisson Processes on the line

4.1 Intervals of a homogeneous process

We have been studiously avoiding the most important example of the state space S, the whole real line $\mathbb{R}$ or a subinterval of it. This has been done to stress that most of the general results have nothing to do with the special structure of $\mathbb{R}$ and are equally applicable to Poisson processes in the plane, in 3-space, or in more general spaces S like those we shall meet in Chapter 7.

The time has now come to link this general theory with the peculiar properties of $\mathbb{R}$, notably its order structure. We consider in detail the homogeneous process, since it was noted in Section 2.1 that most non-homogeneous processes could be made uniform by a simple transformation (cf. Section 4.5).

Suppose then that Π is a Poisson process in $\mathbb{R}$ whose mean measure is λ times Lebesgue measure; the number of points of Π in a Borel set A has the Poisson distribution $\mathscr{P}(\lambda|A|)$. In particular, the number $N(a, b)$ in an interval (a, b) has expectation $\lambda(b - a)$. We speak of Π as a Poisson process with constant rate λ.

The number of points of Π in any bounded set is then finite with probability 1, and Π has no finite limit points. On the other hand, the number in $(0, \infty)$ is infinite, so that the points in $(0, \infty)$ can be written in order as

$$0 < X_1 < X_2 < X_3 < \cdots. \tag{4.1}$$

Similarly the points in $(-\infty, 0)$ can be written in order as

$$\cdots X_{-3} < X_{-2} < X_{-1} < 0. \tag{4.2}$$

These exhaust the points of Π, since $\mathbb{P}\{0 \in \Pi\} = 0$.

It is an important, if almost obvious, fact that the X_n are random variables. For $n \geqslant 1$, for instance, the assertion that $X_n \leqslant x$ is (for any $x > 0$) equivalent to the assertion that $N(0, x] \geqslant n$. In the notation of Section 1.3,

$$\{\omega; X_n(\omega) \leqslant x\} = \{\omega; N(0, x] \geqslant n\} \in \mathscr{F} \tag{4.3}$$

which shows that X_n is a random variable. A similar argument shows that X_{-n} is also a random variable, and since

$$\Pi = \{\ldots, X_{-2}, X_{-1}, X_1, X_2, \ldots\} \tag{4.4}$$

an understanding of the random *sequence* $(X_n; n \neq 0)$ leads at once to an understanding of the random *set* Π.

Because the points of $\Pi_n(-\infty, 0)$ and $\Pi \cap (0, \infty)$ form independent Poisson processes on $(-\infty, 0)$ and $(0, \infty)$ respectively, the subsequences $(X_n; n \leqslant -1)$ and $(X_n; n \geqslant 1)$ are independent. Applying the mapping theorem with $f(x) = -x$ shows that the joint distributions of the sequence

$$(-X_{-n}; n = 0, 1, 2, \ldots)$$

are the same as those of

$$(X_n; n = 1, 2, \ldots)$$

Thus (4.4) is obtained by taking two independent copies of (4.1), one reversed to lie in $(-\infty, 0)$. For this reason it is sufficient (and usual) to consider only the sequence (4.1). And of course there are applications, particularly those in which $\mathbb{R}$ represents a time axis, in which only the positive values are relevant anyway. It should however be noted that such a strong role for the origin weakens one aspect of the homogeneous process, its invariance under translation.

The structure of the random sequence (4.1) is very well known, and particularly simple. The following theorem would merit being called the fundamental theorem of Poisson theory, were it not for the fact that it has no natural analogue when the state space S is of higher dimension.

Interval Theorem *Let Π be a Poisson process of constant rate λ on $(0, \infty)$ and let the points of Π be written in ascending order as in (4.1). Then the random variables*

$$Y_1 = X_1, \qquad Y_n = X_n - X_{n-1} \quad (n \geqslant 2) \tag{4.5}$$

are independent, and each has probability density

$$g(y) = \lambda \, \mathrm{e}^{-\lambda y} \quad (y > 0). \tag{4.6}$$

Proof It is possible, but hard and unrewarding work to use (4.3) to compute the joint distribution of $X_1, X_2, \ldots, X_n$ for each n. A better proof relies on a version of the 'strong Markov property', that if the origin is moved to the random point X_1 then the points of Π to the right of X_1 form a Poisson process independent of X_1.

More precisely, consider the random subset of $(0, \infty)$ defined by

$$\Pi' = \{X_2 - X_1, X_3 - X_1, X_4 - X_1, \ldots\}. \tag{4.7}$$

We wish to prove that X_1 and Π' are independent, and that Π' is a Poisson process of constant rate λ.

To do this, we use the characteristic functional as in Section 3.3. Let f be a continuous function on $(0, \infty)$ which vanishes outside a bounded set and satisfies $f(0) = 0$, and consider the sum

$$\Sigma' = \sum_{n=2}^{\infty} f(X_n - X_1). \tag{4.8}$$

If we can prove that, for each such choice of f, X_1 and Σ' are independent, and that Σ' has the same distribution as

$$\Sigma = \sum_{n=1}^{\infty} f(X_n), \tag{4.9}$$

then it follows that Π' is independent of X_1 and has the same stochastic structure as Π.

Denote by ξ_k the least integer multiple of 2^{-k} greater than X_1. Then ξ_k is a random variable that decreases to X_1 as $k \to \infty$, so that

$$\Sigma' = \lim_{k \to \infty} \Sigma^k, \qquad \Sigma^k = \sum_{n=2}^{\infty} f(X_n - \xi_k), \tag{4.10}$$

with the convention that $f(x) = 0$ if $x < 0$. For any z, x,

$$\mathbb{P}\{\Sigma^k \leqslant z, \quad X_1 \leqslant x\} = \sum_{l=1}^{\infty} \mathbb{P}\{\Sigma^k \leqslant z, \quad X_1 \leqslant x, \quad \xi_k = l2^{-k}\}. \tag{4.11}$$

When $\xi_k = l2^{-k}$ the points X_n in $(l2^{-k}, \infty)$ form a Poisson process on that interval, which is independent of $X_1 < l2^{-k}$. Hence

$$\Sigma^k = \sum_{n=2}^{\infty} f(X_n - l2^{-k})$$

has the same distribution as Σ, and

$$\mathbb{P}\{\Sigma^k \leqslant z, \quad X_1 \leqslant x, \quad \xi_k = l2^{-k}\} = \mathbb{P}\{\Sigma \leqslant z\}\mathbb{P}\{X_1 \leqslant x, \quad \xi_k = l2^{-k}\}.$$

Substituting into (4.11),

$$\begin{aligned}\mathbb{P}\{\Sigma^k \leqslant z, X_1 \leqslant x\} &= \mathbb{P}\{\Sigma \leqslant z\}\mathbb{P}\{X_1 \leqslant x\} \\ &= \mathbb{P}\{\Sigma \leqslant z\}\mathbb{P}\{N(0, x] \geqslant 1\} \\ &= \mathbb{P}\{\Sigma \leqslant z\}(1 - \mathrm{e}^{-\lambda x}).\end{aligned}$$

Letting $k \to \infty$,

$$\mathbb{P}\{\Sigma' \leqslant z, \quad X_1 \leqslant x\} = \mathbb{P}\{\Sigma \leqslant z\} \int_0^x \lambda\, \mathrm{e}^{-\lambda y}\, \mathrm{d}y,$$

which shows that Σ' is independent of X_1 and has the same distribution as Σ, and that X_1 $(= Y_1)$ has the probability density (4.6).

Repeating this argument proves by induction that, for any m, $Y_1, Y_2, \ldots, Y_m$

and Π^m are independent, where

$$\Pi^m = \{X_{m+1} - X_m, X_{m+2} - X_m, \ldots,\}$$

and that the Y_r have the probability density (4.6). This completes the proof.

In this argument, essential use has been made of the ordering of $\mathbb{R}$. For instance, it is not true that, for the bilateral Poisson process (4.4), the points

$$X_n - X_1 \quad (n \neq 1)$$

form a Poisson process. If they did, then $L = X_1 - X_{-1}$ would have the density (4.6), whereas in fact L, the length of the interval containing the origin, has the same density as X_2, namely

$$g_2(y) = \int_0^y g(u)\, g(y-u)\, \mathrm{d}u = \lambda^2 y\, \mathrm{e}^{-\lambda y}. \tag{4.12}$$

This is an example of the *waiting time paradox*. For any x, the distance from x to the next point of Π to the right, and the distance to the next on the left, are independent with density (4.6). Their sum, the length L_x of the interval containing x has the different density (4.12). The point is that this interval is not chosen 'at random' from the intervals of Π. Longer intervals are more likely to be selected than shorter ones, and this sampling bias is exactly expressed by the fact that

$$g_2(y) = \lambda y g(y). \tag{4.13}$$

Equation (4.13) is an example of a very general result in the theory of point processes; see for example Section 5.4 of Cox (1962).

4.2 The Law of Large Numbers

By the theorem of the last section, the nth point X_n of the homogeneous Poisson process on $(0, \infty)$ may be written as the sum

$$X_n = Y_1 + Y_2 + \cdots + Y_n \tag{4.14}$$

of independent random variables, each with the exponential distribution (4.6). Thus the classical theory of sums of independent random variables applies, to give both exact and asymptotic information about X_n.

The distribution of X_n is easily computed. For $x > 0$,

$$\begin{aligned}\mathbb{P}\{X_n \leqslant x\} &= \mathbb{P}\{N(0, x] \geqslant n\} \\ &= \sum_{r=n}^{\infty} \pi_r(\lambda x) \\ &= \int_0^x \lambda \pi_{n-1}(\lambda u)\, \mathrm{d}u,\end{aligned}$$

using (1.12). Hence X_n has probability density

$$g_n(x) = \lambda\pi_{n-1}(\lambda x) = \frac{\lambda^n x^{n-1} e^{-\lambda x}}{(n-1)!}, \tag{4.15}$$

as can also be proved quite easily by induction on n. It follows at once that

$$\mathbb{E}(X_n) = n\lambda^{-1}, \qquad \text{var}(X_n) = n\lambda^{-2}. \tag{4.16}$$

The strong law of large numbers show that, with probability 1,

$$\lim_{n\to\infty} X_n/n = \lambda^{-1}. \tag{4.17}$$

Since

$$N(0, t] \to \infty \qquad \text{as } t \to \infty,$$

it follows that

$$\lim_{t\to\infty} X_{N(0,t]}/N(0, t] = \lambda^{-1},$$

and since

$$X_{N(0,t]} \leqslant t < X_{N(0,t]+1}$$

this means that

$$\lim_{t\to\infty} t/N(0, t] = \lambda^{-1}.$$

Thus we have the strong law of large numbers for the homogeneous Poisson process in the form

$$\lim_{t\to\infty} N(0, t]/t = \lambda,$$

with probability 1.

This simple result is of the first importance, not only for homogeneous processes but, as we shall see in Section 4.5, for non-homogeneous Poisson processes too. It therefore merits a direct proof, not depending on the Interval Theorem.

Law of Large Numbers *Let* Π *be a Poisson process of constant rate* λ *on* $(0, \infty)$. *Then the number* $N(0, t]$ *of points of* Π *in* $(0, t]$ *satisfies*

$$\lim_{t\to\infty} N(0, t]/t = \lambda, \tag{4.18}$$

with probability 1.

Proof The random variable $N(0, t]$ has distribution $\mathscr{P}(\lambda t)$, so that

$$\mathbb{E}\{N(0, t]\} = \lambda t, \qquad \text{var}\{N(0, t]\} = \lambda t. \tag{4.19}$$

Tchebychev's inequality shows that, for any $\varepsilon > 0$,

$$\mathbb{P}\left\{\left|\frac{N(0,t]}{t} - \lambda\right| \geqslant \varepsilon\right\} \leqslant \frac{\lambda}{\varepsilon^2 t}. \tag{4.20}$$

Now take $t_k = k^2$; then (4.20) implies that

$$\sum_{k=1}^{\infty} \mathbb{P}\left\{\left|\frac{N(0,k^2]}{k^2} - \lambda\right| \geqslant \varepsilon\right\} < \infty,$$

so that, by the Borel–Cantelli lemma, there is probability 1 that

$$\left|\frac{N(0,k^2]}{k^2} - \lambda\right| \geqslant \varepsilon$$

for at most a finite number of integer values of k. Since $\varepsilon > 0$ is arbitrary, this shows that

$$\lim_{k\to\infty} \frac{N(0,k^2]}{k^2} = \lambda \tag{4.21}$$

with probability 1.

To get from (4.21) to (4.18) we take k to be the integer part of $t^{1/2}$, so that, for $t > 1$,

$$N(0,k^2] \leqslant N(0,t] \leqslant N(0,(k+1)^2]$$

and

$$k^2 \leqslant t < (k+1)^2.$$

Since

$$(k+1)^2/k^2 \to 1,$$

the proof is complete.

One advantage of this method of proof is that it extends at once to Poisson processes in more general spaces, where the Interval Theorem is not available. Suppose for instance that Π is a Poisson process of constant rate λ on $\mathbb{R}^d$. Let A_k be the sphere of radius k with centre at the origin, so that $N(A_k)$ has expectation and variance both equal to γk^d, where $\gamma = |A_1|$ is a constant depending only on d. For $d \geqslant 2$ the Borel–Cantelli lemma shows that

$$\lim_{k\to\infty} N(A_k)/|A_k| = \lambda, \tag{4.22}$$

with probability 1.

For any bounded set B, let $k(B)$ be the largest integer with $A_k \subseteq B$, and let $K(B)$ be the smallest integer with $B \subseteq A_K$. Then

$$N(A_{k(B)}) \leqslant N(B) \leqslant N(A_{K(B)}),$$

$$|A_{k(B)}| \leqslant |B| \leqslant |A_{K(B)}|,$$

and it follows that

$$\lim \frac{N(B)}{|B|} = \lambda \tag{4.23}$$

with probability 1 so long as

$$k(B) \to \infty, \qquad K(B)/k(B) \to 1. \tag{4.24}$$

More precisely, with probability 1 there exist for every $\varepsilon > 0$ numbers N, δ such that, for any set B with

$$k(B) \geqslant N, \qquad K(B)/k(B) \leqslant 1 + \delta, \tag{4.25}$$

we have

$$\left|\frac{N(B)}{|B|} - \lambda\right| < \varepsilon. \tag{4.26}$$

4.3 Queues

We have remarked that, in many of the applications of Poisson process theory in one dimension, the points of Π represent random instants of time, like the instants at which a piece of radioactive material emits a particle recorded by a Geiger counter.

Another such example is that of arrivals at some sort of service facility. The theory of queues, a major technique of operational research, grew out of the observation of telephone traffic engineers (notably A. K. Erlang in Copenhagen) that the times at which calls arrive at a telephone exchange are well described by a Poisson process of constant, or slowly varying rate.

It is not the object of this section to describe all the ramifications of the theory of queues. The interested reader may like to consult the introductory text by Cox and Smith (1961) or the excellent modern account of Asmussen (1987). All that is attempted here is an illustration of the way in which the general theory of Poisson processes informs the analysis of simple queueing models.

Consider first the queue conventionally denoted by the symbol $M/M/1$. This is a single-server queue at which customers arrive according to a Poisson process Π_α of constant rate α. They queue in order of arrival for the attentions of the server, who looks after the needs of each in turn in such a way that the service times of different customers are independent and have the probability density

$$\beta\, \mathrm{e}^{-\beta t} \quad (t > 0). \tag{4.27}$$

This particular service time distribution is very easy to deal with, because it allows a representation in terms of a second, independent Poisson process. Thus suppose that Π_β is a Poisson process independent of Π_α and of constant rate β. Instruct the server to go on serving the customer at the head of the queue until the next instant of Π_β, at which point that customer leaves and the next customer (if any) enters service. Then it is an immediate consequence of the Interval Theorem that the resulting service times are indeed independent with density (4.27). The points of Π_β are then called *potential service instants.* Every instant at which a customer completes service and leaves is a point of Π_β, but not conversely; a point of Π_β that falls when the queue is empty will not be an actual service instant.

The traditional way of analysing the queue $M/M/1$ is to formulate the queue length process as a continuous-time Markov process, and to write down and solve the Kolmogorov forward differential equations. Thus let $p_n(t)$ be the probability that at time t there are n customers present. For small $h > 0$, the change from $p_n(t)$ to $p_n(t + h)$ must take account of the fact that, in the interval $(t, t + h)$, there will be an arrival with probability $\alpha h + o(h)$ and (if $n \geqslant 1$) a departure with probability $\beta h + o(h)$. Thus

$$p_n(t + h) = p_n(t)(1 - \alpha h - \beta h + o(h)) \\ + p_{n-1}(t)(\alpha h + o(h)) + p_{n+1}(t)(\beta h + o(h)) + o(h)$$

leading as $h \to 0$ to the differential equation

$$\frac{\mathrm{d}p_n}{\mathrm{d}t} = -(\alpha + \beta)p_n(t) + \alpha p_{n-1}(t) + \beta p_{n+1}(t), \quad (n \geqslant 1). \tag{4.28}$$

When $n = 0$ the same argument leads to

$$\frac{\mathrm{d}p_o}{\mathrm{d}t} = -\alpha p_0(t) + \beta p_1(t). \tag{4.29}$$

This family of simultaneous differential equations is most easily solved by Laplace transforms. Suppose for definiteness that the queue is empty when $t = 0$, so that

$$p_n(0) = \delta_{n0}. \tag{4.30}$$

Writing

$$r_n(\lambda) = \int_0^\infty p_n(t)\,\mathrm{e}^{-\lambda t}\,\mathrm{d}t, \tag{4.31}$$

(4.28) and (4.30) show that, for $n \geqslant 1$,

$$\lambda r_n(\lambda) = -(\alpha + \beta)r_n(\lambda) + \alpha r_{n-1}(\lambda) + \beta r_{n+1}(\lambda).$$

This implies that, for fixed λ, r_n is a linear combination of ξ^n and η^n, where ξ and η are the roots of the quadratic

$$\beta x^2 - (\alpha + \beta + \lambda)x + \alpha = 0. \tag{4.32}$$

It is easy to check that exactly one of the roots (ξ, say) lies in $|x| \leqslant 1$, and since r_n is bounded it must be a multiple of ξ^n. Since

$$\sum_{n=0}^{\infty} r_n(\lambda) = \int_0^{\infty} \sum_{n=0}^{\infty} p_n(t)\, \mathrm{e}^{-\lambda t}\, \mathrm{d}t = \int_0^{\infty} \mathrm{e}^{-\lambda t}\, \mathrm{d}t = \lambda^{-1},$$

we must have

$$r_n(\lambda) = \lambda^{-1}(1 - \xi)\, \xi^n, \tag{4.33}$$

where

$$\xi = \xi(\lambda) = \frac{1}{2\beta}\{\alpha + \beta + \lambda - [(\alpha + \beta + \lambda)^2 - 4\alpha\beta]^{1/2}\}. \tag{4.34}$$

In particular, $p_0(t)$ has Laplace transform

$$r_0(\lambda) = \frac{1}{2\beta\lambda}\{\beta - \alpha - \lambda + [(\alpha + \beta + \lambda)^2 - 4\alpha\beta]^{1/2}\}, \tag{4.35}$$

which has a standard (but complicated and uninformative) inversion in terms of Bessel functions. Much more interesting, however is the following formula in terms of the Poisson probabilities

$$p_0(t) = \sum_{0 \leqslant r \leqslant s} \frac{s - r + 1}{s + 1} \pi_r(\alpha t) \pi_s(\beta t). \tag{4.36}$$

This can be verified by computing the Laplace transform $r(\lambda)$ of the right-hand side of (4.36), and comparing it with the formula (4.35) for $r_0(\lambda)$. Thus

$$\begin{aligned} r(\lambda) &= \int_0^{\infty} \sum_{0 \leqslant r \leqslant s} \frac{s - r + 1}{(s + 1) r!\, s!} (\alpha t)^r (\beta t)^s\, \mathrm{e}^{-(\alpha + \beta + \lambda)t}\, \mathrm{d}t \\ &= \sum_{0 \leqslant r \leqslant s} \frac{(s - r + 1)(r + s)!}{r!\,(s + 1)!} \frac{\alpha^r \beta^s}{(\alpha + \beta + \lambda)^{r + s + 1}}. \end{aligned}$$

Writing

$$u = \alpha/(\alpha + \beta + \lambda), \qquad v = \beta/(\alpha + \beta + \lambda),$$

$$\lambda r(\lambda) = (1 - u - v) \sum_{0 \leqslant r \leqslant s} \frac{(s - r + 1)(r + s)!}{r!\,(s + 1)!} u^r v^s.$$

If the coefficient of $u^r v^s$ on the right-hand side is computed and simplified, it will be found to be zero unless $r = s = 0$ (when it is 1) or $s = r - 1$, when it is equal to

$$-\frac{(2r - 2)!}{(r - 1)!\, r!}$$

Thus

$$\lambda r(\lambda) = 1 - \sum_{r=1}^{\infty} \frac{(2r-2)!}{(r-1)!\, r!} u^r v^{r-1}$$

$$= 1 - \frac{1}{2v}\{1 - (1 - 4uv)^{1/2}\}$$

$$= 1 - \xi = \lambda r_0(\lambda),$$

proving (4.36).

Equation (4.36) cries out for an interpretation in terms of a Bernoulli process of arrivals. Since (4.36) holds for all α, the coefficient of $\pi_r(\alpha t)$ is the probability that the queue is empty at time t when the arrivals form, not a Poisson process but a Bernoulli process (Section 2.4) in which exactly r customers arrive independently according to a uniform distribution on $(0, t)$:

$$p_0(r, t) = \sum_{s=r}^{\infty} \frac{s - r + 1}{s + 1} \pi_s(\beta t). \tag{4.37}$$

This argument can be repeated. The coefficient of $\pi_s(\beta t)$ is the probability of an empty queue at time t if the s potential service instants form a Bernoulli process on $(0, t)$:

$$p_0(r, s; t) = \frac{s - r + 1}{s + 1} \quad (r \leqslant s). \tag{4.38}$$

This is now a purely combinatorial result. If r arrival instants and s potential service instants are independently and uniformly distributed on an interval (whose length is now irrelevant) the probability that there are no customers at the end of the interval is given by (4.38). This is an example of a *ballot theorem*, and the connection between such results and the theory of queues has been the subject of much study, noticeably by Takács (1967). Our argument can be reversed; if a direct combinatorial proof can be found (and Takács shows how this can be done), then (4.36) can be proved without the need for differential equations or Laplace transforms.

4.4 Bartlett's Theorem

The queue $M/M/1$ is only the simplest of many such systems, some of great complexity, and the argument leading to (4.28) can be greatly generalised. For instance, the system $M/M/k$ differs only in having k servers. A customer only needs to be served by one of them, and a server on becoming free attends to the foremost customer not already being served. Equations (4.28) and (4.29) are then easily seen to be replaced by

$$\begin{aligned}\frac{dp_n}{dt} &= -(\alpha + \beta k)p_n + \alpha p_{n-1} + \beta k p_{n+1} \quad (n \geqslant k),\\ \frac{dp_n}{dt} &= -(\alpha + \beta n)p_n + \alpha p_{n-1} + \beta(n+1)p_{n+1} \quad (n < k),\end{aligned} \tag{4.39}$$

The general solution of these is complicated, but in many cases the system is in statistical equilibrium, so that $p_n = p_n(t)$ is independent of t. Then (4.39) becomes

$$(\alpha + \beta n \wedge k)p_n = \alpha p_{n-1} + \beta(n+1) \wedge k p_{n-1},$$

where $n \wedge k$ is the lesser of n and k. Thus

$$\alpha p_n - \beta(n+1) \wedge k p_{n+1} = \alpha p_{n-1} - \beta n \wedge k p_n,$$

showing that

$$\alpha p_{n-1} - \beta n \wedge k p_n$$

is a constant, which must be zero since $p_n \to 0$ $(n \to \infty)$. Thus

$$p_n/p_{n-1} = \alpha/(n \wedge k)\beta,$$

whence

$$\begin{aligned} p_n &= p_0\left(\frac{\alpha}{\beta}\right)^n \frac{1}{n!} \quad (n \leqslant k),\\ p_n &= p_0\left(\frac{\alpha}{\beta}\right)^n \frac{1}{k!\,k^{n-k}} \quad (n > k),\end{aligned} \tag{4.40}$$

where p_0 is determined by the equation

$$1 = \sum_{n=0}^{\infty} p_n = p_0\left\{\sum_{n=0}^{k}\left(\frac{\alpha}{\beta}\right)^n \frac{1}{n!} + \sum_{n=k+1}^{\infty}\left(\frac{\alpha}{\beta}\right)^n \frac{1}{k!\,k^{n-k}}\right\}. \tag{4.41}$$

This works if the final series in (4.41) converges, which it does if $\alpha < \beta k$. Otherwise there is no stationary solution of (4.39) and the queue is unstable; it can then be shown that, whatever the initial condition in (4.39),

$$p_n(t) \to 0 \quad (t \to \infty)$$

for all n. (It is an interesting exercise to verify this directly when $k = 1$, from (4.37).)

Now suppose that $k = \infty$, so that there is an unlimited supply of servers. Then (4.40) shows that

$$p_n = p_0\left(\frac{\alpha}{\beta}\right)^n \frac{1}{n!},$$

(4.41) shows that $p_0 = e^{-\alpha/\beta}$ and so the number of customers present, in equilibrium, has the Poisson distribution $\mathscr{P}(\alpha/\beta)$. But in fact even more is true. The equations (4.39) when $k = \infty$ take the form

$$\frac{\mathrm{d}p_n}{\mathrm{d}t} = -(\alpha + \beta n)p_n + \alpha p_{n-1} + \beta(n+1)p_{n+1}, \tag{4.42}$$

and these admit a whole family of solutions of the form $\mathscr{P}(\mu(t))$. If we substitute $p_n(t) = \pi_n(\mu(t))$ in (4.42), it becomes

$$\frac{\mathrm{d}\mu}{\mathrm{d}t}(\pi_{n-1} - \pi_n) = -\alpha\pi_n - \beta\mu\pi_{n-1} + \alpha\pi_{n-1} + \beta\mu\pi_n,$$

which is satisfied for all n so long as

$$\frac{\mathrm{d}\mu}{\mathrm{d}t} = \alpha - \beta\mu. \tag{4.43}$$

The general solution of this equation is

$$\mu(t) = \frac{\alpha}{\beta} + \left(\mu(0) - \frac{\alpha}{\beta}\right)\mathrm{e}^{-\beta t}. \tag{4.44}$$

Thus, if the number in the queue at $t = 0$ has a Poisson distribution, so does the number at any time $t > 0$, the mean being given by (4.44).

This is a very striking result, which demands a probabilistic explanation. This was given by Bartlett (1949), and shows that the features of the queue $M/M/\infty$ which imply the persistence of the Poisson distribution are, that the customers arrive in a Poisson process, and that while in the system they do not interfere with one another (for instance by competing for a limited number of servers).

We express Bartlett's result in terms of 'customers' arriving at a 'system', but these words are arbitrary labels. The customers could be machines breaking down and 'queueing' for repair, or radioactive particles falling on a counter, or seeds falling from a tree and entering a series of biological phases.

Bartlett's Theorem *Customers arrive according to a Poisson process of rate $\alpha(t)$. They then move at random around the system in such a way that their trajectories are independent. Let E be a subset such that the probability of a customer who arrives at time s being in E at a subsequent time t is $p(s, t)$. Then the number of customers in E at time t has a Poisson distribution with mean*

$$\mu(t) = \int_{-\infty}^{t} \alpha(s)p(s, t)\,\mathrm{d}s. \tag{4.45}$$

It is not difficult to prove this result directly, but it is an almost trivial consequence of a theorem to be proved in the next chapter. We therefore defer the proof of Section 5.2, and simply note here the connection between (4.44) and (4.45). For simplicity suppose that customers start arriving at time 0 so that

$$\alpha(t) = 0 \quad (t < 0), \qquad \alpha(t) = \alpha \quad (t > 0). \tag{4.46}$$

Clearly

$$p(s, t) = q(t - s), \tag{4.47}$$

where $q(t)$ is the probability that the customer spends more than time t in the queue. Then (4.45) becomes

$$\mu(t) = \int_0^t \alpha q(t - s)\, \mathrm{d}s = \alpha \int_0^t q(u)\, \mathrm{d}u, \tag{4.48}$$

and since

$$q(t) = \int_t^\infty \beta\, \mathrm{e}^{-\beta s}\, \mathrm{d}s = \mathrm{e}^{-\beta t},$$

(4.48) becomes

$$\mu(t) = \frac{\alpha}{\beta}(1 - \mathrm{e}^{-\beta t}), \tag{4.49}$$

which is (4.45) with $\mu(0) = 0$.

More generally, the queue $M/G/\infty$, in which the service time has a general distribution with distribution function $B(s)$, satisfies (4.48) with $q(t) = 1 - B(t)$, so that

$$\mu(t) = \alpha \int_0^t \{1 - B(u)\}\, \mathrm{d}u. \tag{4.50}$$

4.5 Non-homogeneous processes

The justification for concentrating, in one dimension, on homogeneous processes is that a non-homogeneous process can be made homogeneous by a monotone transformation. More precisely, consider a Poisson process Π on $\mathbb{R}$ whose mean measure μ is finite on bounded intervals. Then the function $M(t)$ defined by (2.11) is continuous and increasing, and

$$\mu(a, b] = M(b) - M(a) \tag{4.51}$$

whenever $a < b$. The function M uniquely determines μ.

It was shown in Section 2.3 that the random set

$$\Pi_1 = M(\Pi) = \{M(X);\, X \in \Pi\} \tag{4.52}$$

is a Poisson process of rate 1 on the (finite or infinite) interval

$$(M(-\infty), M(\infty)), \tag{4.53}$$

where

$$M(\pm\infty) = \lim_{t \to \pm\infty} M(t).$$

All the theorems of this chapter apply to Π_1 and hence imply properties of Π by inversion of M; the only possible ambiguity arises if M has flat sections for which $M^{-1}\{y\}$ is an interval rather than a single point, but this can happen at most for countably many points y and the probability that Π_1 contains such a point is zero.

Unfortunately this non-linear transformation of the line destroys some of the most useful properties. In particular, although the intervals of Π_1 are independent, those of Π are not. For this reason it is usually easiest if the process is transformed to homogeneity before any serious analysis is attempted.

To take a concrete example, suppose that Π_2 is a Poisson process in the plane with constant rate λ and let Π be the set of distances from the origin of the points of Π_2. Then it was shown in Section 2.3 that Π is a Poisson process with rate $\lambda(x) = 2\pi\lambda x$ on $(0, \infty)$, so that

$$M(x) = \pi\lambda x^2 \quad (x > 0).$$

This shows that the *squared* distances are more useful than the distances themselves, and form a Poisson process with constant rate $\pi\lambda$.

The strong law of large numbers translates easily into a result about non-homogeneous processes. Suppose that $M(\infty) = \infty$, so that the points

$$M(X) \quad (X \in \Pi, X > 0)$$

form a Poisson process of rate 1 on $(0, \infty)$. Enumerate the points as $0 < Y_1 < Y_2 < \cdots$. By (4.17),

$$\lim_{n \to \infty} Y_n/n = 1 \tag{4.54}$$

with probability 1. The number $N(0, t]$ of points of Π in $(0, t]$ equals the number of points Y_n in $(0, \mathrm{M}(t)]$, and (4.54) then implies that

$$\lim_{t \to \infty} N(0, t]/M(t) = 1. \tag{4.55}$$

Thus the points of Π in $(0, \infty)$ form an infinite sequence without finite limit point, whose asymptotic behaviour is, with probability 1, governed by (4.55).

A similar result is true in $(-\infty, 0)$, if $M(-\infty) = -\infty$. If $M(-\infty)$ is finite, there are only a finite number of points of Π to the left of the origin.

Not all interesting processes have the property that μ is finite on bounded intervals. We shall for instance meet in Chapters 8 and 9 a Poisson process with rate $\lambda(x)$ for which

$$\int_0^\infty \lambda(x)\,\mathrm{d}x = \infty, \qquad \int_t^\infty \lambda(x)\,\mathrm{d}x < \infty \quad (t > 0). \tag{4.56}$$

It is then natural to consider not M but L, defined by

$$L(t) = \int_t^\infty \lambda(x)\,\mathrm{d}x. \tag{4.57}$$

Then L is continuous and decreasing on $(0, \infty)$, with

$$L(0) = \infty, \qquad L(\infty) = 0. \tag{4.58}$$

Exactly as in Section 2.3, the function L maps the process Π into a Poisson process Π_1 of rate 1 on $(0, \infty)$. There are infinitely many points of Π_1, with no finite limit point. Inverting the mapping, it follows that there are infinitely many points of Π in the neighbourhood of the origin, but only finitely many to the right of any $t > 0$.

In this case the law of large numbers gives information about the clustering of points of Π near the origin. The number $N(t, \infty)$ of points to the right of t equals the number of points of $\Pi_1 = L(\Pi)$ to the left of $L(t)$. Hence, with probability 1,

$$\lim_{t \to \infty} N(t, \infty)/L(t) = 1. \tag{4.59}$$

This apparently arcane result has applications as diverse as population genetics and the design of dams.

5

Marked Poisson processes

5.1 Colouring

It was noted in Section 1.2 that the Poisson distribution has a characteristic property in relation to the binomial distribution. If the random variable N has distribution $\mathscr{P}(\mu)$ and if a second random variable M has a conditional distribution, given N, of the form $\mathscr{B}(N, p)$, then M and $N - M$ are independent Poisson variables, with respective means $p\mu$ and $(1 - p)\mu$.

This has an immediate, and surprising, consequence for Poisson processes. Let Π be a Poisson process on the general state space S, with mean measure μ. Colour the points of Π randomly either red or green, where the colours of different points are independent, the respective probabilities of red and green being p and $q = 1 - p$.

For any $A \subseteq S$, write $N(A)$, $N_r(A)$, $N_g(A)$, for the number of points of Π in A, the number of red points, and the number of green points. Then $N(A)$ has distribution $\mathscr{P}(\mu(A))$ and, given $N(A)$, the conditional distribution of $N_r(A)$ is $\mathscr{B}(B(A), p)$. Hence $N_r(A)$ and $N(A) - N_r(A) = N_g(A)$ are independent, with distributions $\mathscr{P}(p\mu(A))$, $\mathscr{P}(q\mu(A))$.

Moreover, if $A_1, A_2, \ldots, A_n$ are disjoint sets in S, the triples of random variables

$$(N(A_j), N_r(A_j), N_g(A_j)), \quad j = 1, 2, \ldots, n,$$

are independent. Hence the $2n$ random variables

$$N_r(A_j). \qquad N_g(A_j), \quad j = 1, 2, \ldots, n,$$

are independent. It follows that the set of red points and the set of green points are independent Poisson processes with respective mean measures $p\mu$, $q\mu$.

This result extends by induction on k to colouring by any finite number k of colours, and we have the following theorem.

Colouring Theorem *Let Π be a Poisson process on S with mean measure μ. Let the points of Π be coloured randomly with k colours, the probability that a point receives the ith colour being p_i and the colours of different points being independent (of one another and of the positions of the points). Let Π_i be the set of points with the ith colour. Then the Π_i are independent Poisson processes with mean measures*

$$\mu_i = p_i\mu. \tag{5.1}$$

Note that this is consistent with the Superposition Theorem, since Π is the superposition of the independent Π_i and

$$\mu = \sum_{i=1}^{k} p_i \mu = \sum_{i=1}^{k} \mu_i.$$

When one comes to apply this theorem, it quickly becomes clear that it is too restrictive. Attempts to generalise it lead to the theory of marked Poisson processes which is the main topic of this chapter.

To see the sort of generalisation to be sought, consider a simple example. Suppose that cars are moving along a long stretch of road. Taking the road as the line $\mathbb{R}$, and representing each car as a point, a snapshot of the road at time t will show a subset of $\mathbb{R}$ (without, for safety's sake, finite limit points). It might be realistic to model this as a Poisson process.

The distribution will change with time because cars are moving at different speeds. One may ask whether, if the cars form a Poisson process at $t = 0$, they still form such a process at later times t.

This question is easy to answer if the speeds of cars are random variables, independent of one another and of position on the road, if they do not change with time and in particular if overtaking is unrestricted, and if they take only a finite number of values. If the possible speeds are $v_1, v_2, \ldots, v_k$, and Π_i denotes the sets of cars of speed v_i at $t = 0$, then the Colouring Theorem shows that the Π_i are independent Poisson processes. At a later time the positions of those cars form the Poisson process $\Pi_i + v_i t$ obtained by translating Π_i by an amount $v_i t$ (and using the Mapping Theorem). The positions of all cars at time t is then the superposition

$$\Pi(t) = \bigcup_{i=1}^{k} (\Pi_i + v_i t), \tag{5.2}$$

which by the Superposition Theorem is again a Poisson process.

The mean measure of $\Pi(t)$ is easily computed. Suppose for simplicity that Π has a rate $\lambda(x)$. Then Π_i has rate $p_i \lambda(x)$, if p_i is the probability of speed v_i, and $\Pi_i + v_i t$ has rate $p_i \lambda(x - v_i t)$. It follows that $\Pi(t)$ has rate

$$\lambda_t(x) = \sum_{i=1}^{k} p_i \lambda(x - v_i t). \tag{5.3}$$

In particular, if $\lambda(x)$ is a constant λ, (5.3) shows that

$$\lambda_t(x) = \lambda \qquad \text{for all } t > 0; \tag{5.4}$$

if the cars form a homogeneous Poisson process at $t = 0$ they form a similar process at all later times.

This engagingly simple result has, however, been proved under very restrictive conditions. The requirement that there be only a finite number of possible speeds can be removed by an approximation argument. More serious is the assumption that the speed distribution is independent of position, and that speeds remain constant.

To get round the difficulty requires a way of 'marking' each point in such a way that the marks are independent of each other but may depend on the positions of the points marked. The marks can be of a very general form, but there is then a powerful generalisation of the Colouring Theorem which deals at once with problems such as that described above.

5.2 The product space representation

The idea is basically very simple, but it requires some care to formulate precisely. Let Π be a Poisson process on S with mean measure μ. Suppose that, with each point X of the random set Π, we associate a random variable m_X (the *mark* of X) taking values in some space M. The distribution of m_X may depend on X but not on the other points of Π, and the m_X for different X are independent.

The pair (X, m_X) can then be regarded as a random point X^* in the product space $S \times M$. The totality of points X^* forms a random countable subset

$$\Pi^* = \{(X, m_X); X \in \Pi\} \tag{5.5}$$

of $S \times M$. The fundamental result, from which all else follows, is that Π^* is a Poisson process on the product space $S \times M$.

The formal definition proceeds as follows. The spaces S and M have the measure-theoretic structure described in Section 2.1. We start with a Poisson process Π on S with mean measure μ, and a probability distribution $p(x, \cdot)$ on M depending on $x \in S$ in such a way that, for $B \subseteq M$, $p(\cdot, B)$ is a measurable function on S. A *marking* of Π is a random subset (5.5) of $S \times M$ whose projection onto S is Π and which is such that the conditional distribution of Π^* given Π makes the m_X independent with respective distributions $p(X, \cdot)$.

Marking Theorem *The random subset Π^* is a Poisson process on $S \times M$ with mean measure μ^* given by*

$$\mu^*(C) = \iint_{(x,m)\in C} \mu(\mathrm{d}x)p(x, \mathrm{d}m). \tag{5.6}$$

Proof This is most easily carried out using the characteristic functional. For any measurable function f on $S \times M$, let

$$\Sigma^* = \sum_{X\in\Pi} f(X, m_X). \tag{5.7}$$

Given Π, Σ^* is the sum of independent random variables $f(X, m_X)$, so that

$$\mathbb{E}\{e^{-\Sigma^*}|\Pi\} = \prod_{X\in\Pi} \mathbb{E}\{e^{-f(X,m_X)}|\Pi\}$$

$$= \prod_{X\in\Pi} \int_M e^{-f(X,m)} p(X, dm).$$

Applying (3.29) to the Poisson process Π with f replaced by

$$f_*(x) = -\log \int_M e^{-f(X,m)} p(x, dm),$$

we have

$$\mathbb{E}\{e^{-\Sigma^*}\} = \exp\left\{-\int_S (1 - e^{-f_*(x)})\mu(dx)\right\}$$

$$= \exp\left\{-\int_S \int_M (1 - e^{-f(x,m)})\mu(dx)p(x, dm)\right\}$$

$$= \exp\left\{-\int_{S\times M} (1 - e^{-f})\, d\mu^*\right\},$$

which shows that Π^* is a Poisson process with mean measure μ^*.

The theorem has a number of immediate but useful corollaries. For instance, the Mapping Theorem shows that, since the points (X, m_X) form a Poisson process on $S \times M$, the marks m_X form a Poisson process on M. The mean measure μ_m is obtained by putting $C = S \times B$ in (5.6):

$$\mu_m(B) = \int_S \int_B \mu(dx)p(x, dm). \tag{5.8}$$

If the marks take on only k different values, the theorem shows that the points with the ith mark form a Poisson process Π_i with mean measure

$$\mu_i(A) = \int_A \mu(dx)p(x, \{m_i\}) \tag{5.9}$$

and that the Π_i $(i = 1, 2, \ldots, k)$ are independent. This generalises the Colouring Theorem by allowing the colouring probabilities $p_i = p(x, \{m_i\})$ to vary with position x.

In particular, the promised proof of Bartlett's Theorem is immediate. In the language of Section 4.4, colour a customer arriving at time s red if he is in E at the subsequent time t. The colours of different customers are independent, and the probability of the arrival at s being red is $p(s, t)$. Hence the arrival times of the red customers form a Poisson process on $(-\infty, t)$ with mean measure

$$\mu(A) = \int_A \alpha(s)p(s,t)\,ds.$$

In particular, the total number of red customers has a Poisson distribution with mean

$$\mu(-\infty, t) = \int_{-\infty}^{t} \alpha(s)p(s,t)\,ds,$$

as claimed.

In some applications it is natural to start with the product space, and this can yield somewhat greater generality. As a simple example, think of a large flat pavement (represented by the plane $\mathbb{R}^2$) when it begins to rain. At first the raindrops form widely separated circles of water on the pavement, but as time goes on, more and more drops arrive and the scattered circles begin to overlap and to form large irregular wet areas.

This can be modelled by means of a Poisson process in $\mathbb{R}^2 \times (0, \infty)$, where the point (X, t_X) denotes a drop that falls at the point X at time t_X. The points X with $t_X \leqslant t$ form a Poisson process $\Pi(t)$ on $\mathbb{R}^2$ which increases as t increases.

If for example the process is uniform in space and time, so that the mean measure on $\mathbb{R}^2 \times (0, \infty)$ has uniform density λ, then $\Pi(t)$ is a homogeneous Poisson process on $\mathbb{R}^2$ with density λt. The process of all points X is uninteresting, since it is infinite on all sets of positive measure; the interesting object is the increasing family of Poisson process $\Pi(t)$.

The process $\Pi(t)$ describes the positions of the centres of the wet circles at time t. It may be reasonable to assume that the radii of these circles are independent random variables, and then we have a marking of $\Pi(t)$ by the radii r_X. The points (X, t_X, r_X) then form a Poisson process in the four-dimensional space $\mathbb{R}^2 \times (0, \infty) \times (0, \infty)$.

5.3 Campbell's Theorem revisited

Campbell's Theorem determines the distribution of the sum $\sum f(x)$ over a Poisson process of points X. Sometimes the quantity to be summed depends not deterministically on X but in some random way, independently from point to point. In other words, we may be interested in the distribution of $\sum m_X$, where m_X is a marking of Π.

For such sums Campbell's Theorem can still be applied, since the m_X form a Poisson process on M. Thus if for example the m_X are positive random variables,

$$\mathbb{E}\{e^{-\sum m_X}\} = \exp\left\{-\int_M (1 - e^{-m})\mu_m(dm)\right\}$$

$$= \exp\left\{-\int_S \int_M (1 - e^{-m})\mu(dx)p(x, dm)\right\}. \quad (5.10)$$

This is a direct generalisation of (3.29), to which it reduces if $m_X = f(X)$, a non-random function of X.

This line of argument can be extended. To illustrate the possibilities, consider a very simple classical model of the universe. Regard the galaxies as point masses, scattered over three-dimensional space in a Poisson process with (variable) density $\lambda(x)$. Suppose that the masses of the galaxies are independent random variables, the mass m_X of a galaxy at X having density $\rho(X, m)$ in $m > 0$. The points (X, m_X) then form a Poisson process Π^* in $\mathbb{R}^3 \times (0, \infty)$ with density

$$\lambda^*(x, m) = \lambda(x)\rho(x, m). \quad (5.11)$$

Now consider the gravitational field at a particular point, the origin for instance. This is a 3-vector (F_1, F_2, F_3), where

$$F_j = \sum_{X \in \Pi} \frac{G m_X X_j}{(X_1^2 + X_2^2 + X_3^2)^{3/2}} \quad (5.12)$$

and G is the gravitational constant.

Applying (3.21) to Π^*, we have

$$\mathbb{E}\{e^{it_1F_1 + it_2F_2 + it_3F_3}\} = \exp\left\{\int_{\mathbb{R}^3} \int_0^\infty (e^{iGmt.\psi(X)} - 1)\lambda(x)\rho(x, m)\, dx\, dm\right\} \quad (5.13)$$

where ψ is the vector function with components

$$\psi_j(x) = (x_1^2 + x_2^2 + x_3^2)^{-3/2} x_j \quad (5.14)$$

and $t.\psi$ is a scalar product. This is valid so long as (3.16) holds, and this is equivalent to

$$\int_{\mathbb{R}^3} m|\psi(x)|\lambda(x)\rho(x, m)\, dx\, dm < \infty. \quad (5.15)$$

If

$$\bar{m}(x) = \int m\rho(x, m)\, dm$$

denotes the expected mass of a galaxy at x, this in turn is the same as the condition

$$\int_{\mathbb{R}^3} \frac{\bar{m}(x)\lambda(x)}{x_1^2 + x_2^2 + x_3^2}\, dx < \infty. \quad (5.16)$$

If (5.16) holds, then (5.13) determines in principle the joint distribution of F_1, F_2, F_3. The means, variances and covariances can be found from (3.18):

$$\mathbb{E}\{F_j\} = \iint Gm\psi_j(x)\lambda(x)\rho(x, m)\,\mathrm{d}x\,\mathrm{d}m$$

$$= G\int_{\mathbb{R}^3} \bar{m}(x)\psi_j(x)\lambda(x)\,\mathrm{d}x, \tag{5.17}$$

and

$$\operatorname{cov}\{F_j, F_k\} = \iint Gm\psi_j(x)Gm\psi_k(x)\lambda(x)\rho(x, m)\,\mathrm{d}x\,\mathrm{d}m$$

$$= G^2\int_{\mathbb{R}^3} \bar{m}_2(x)\psi_j(x)\psi_k(x)\lambda(x)\,\mathrm{d}x, \tag{5.18}$$

where

$$\bar{m}_2(x) = \int_0^\infty m^2\rho(x, m)\,\mathrm{d}m. \tag{5.19}$$

If the universe is uniform, in the sense that $\bar{m}(x)$ and $\lambda(x)$ are non-zero constants, then (5.16) does not hold, so that the sum (5.12) is almost surely not absolutely convergent. This is the gravitational version of Olbers' paradox. Nor does (5.17) converge, but it is a curious fact that (5.18) does converge, if one excludes a neighbourhood of the origin. Thus it is possible to study the variances and covariances in a formal way, starting with (5.18) and ignoring the divergence of (5.12). Of such 'renormalisation' procedures is the stuff of theoretical physics.

5.4 The wide motorway

We now return to the road traffic example of Section 5.1. Suppose that at time $t = 0$ the cars are represented by the points of a Poisson process Π on $\mathbb{R}$ with rate function $\lambda(x)$. We wish to describe the array of cars at a later time $t > 0$, making minimal assumptions about the way they move between 0 and t.

In fact it is only necessary to assume that the movements of different cars are independent. This is true in practice only on wide roads where overtaking is uninhibited, whence the title of this section. In such situations, the Marking and Mapping Theorems show at once that the cars form a Poisson process $\Pi(t)$ at each later time t.

More precisely, denote by $Y_t(X)$ the position at time t of a car which was at X at time 0. If the $Y_t(X)$ for different X are independent, the $Y_t(X)$ form a marking of Π. By the Marking Theorem, the points $(X, Y_t(X))$ form a Poisson process Π^* on $\mathbb{R}^2$. By the Mapping Theorem, the points $Y_t(X)$ form a Poisson process $\Pi(t)$ on $\mathbb{R}$, as asserted.

It is important to note that t has been held fixed throughout this argument. There is no assertion that the $\Pi(t)$ for different t are (say) independent, and in fact the structure of their dependence is quite complicated. If it is necessary to examine the joint distributions of $\Pi(t)$ for a number of different t, this is best done by using the Marking Theorem with $M = \mathbb{R}^n$ to deduce that

$$\{(X, Y_{t_1}(X), Y_{t_2}(X), \ldots, Y_{t_n}(X)); X \in \Pi\} \tag{5.20}$$

is a Poisson process on $\mathbb{R}^{n+1}$, for any $t_1, t_2, \ldots, t_n$. Or more ambitiously, that

$$\{(X, Y_{\cdot}(X)); X \in \Pi\} \tag{5.21}$$

is a Poisson process on $\mathbb{R} \times M$, where the mark space M is a suitable function space of possible trajectories.

The mean measure of $\Pi(t)$ is easily computed. Suppose for simplicity that the conditional distribution of $Y_t(X)$ given X has a probability density $\rho_t(X, \cdot)$. Then by (5.5) the process Π^* on $\mathbb{R}^2$ has rate $\lambda(x)\rho_t(x, y)$. Hence

$$\lambda_t(y) = \int_{-\infty}^{\infty} \lambda(x)\rho_t(x, y)\,\mathrm{d}x. \tag{5.22}$$

Suppose for example that the cars have constant velocities drawn independently from a velocity distribution with probability density $g(v)$. Then

$$\rho_t(x, y) = t^{-1} g((y - x)/t), \tag{5.23}$$

so that (5.22) becomes

$$\lambda_t(y) = \int_0^{\infty} \lambda(y - vt) g(v)\,\mathrm{d}v. \tag{5.24}$$

Note that, if $\lambda(x) = \lambda$, then $\lambda_t(y) = \lambda$ for all t; a homogeneous Poisson process is preserved under independent velocities. This simple conclusion depends however on the special structure of (5.23), and will not generally be true of (5.22).

It is important to emphasise that these simple results depend crucially on the assumption that the cars move independently. Once they are allowed to interfere with one another, the Poisson character is lost. Cars form 'platoons' from which the faster cars can only escape when they find an opportunity to overtake; they then overhaul the next platoon, and the cycle of frustration is repeated.

5.5 Ecological models

There is nothing especially one-dimensional about the analysis of the previous section, and there are other applications in which a state space of higher dimension is appropriate. It is worth restating some of the conclusions in this more general context.

Displacement Theorem *Let Π be a Poisson process on $\mathbb{R}^d$ with rate function $\lambda(x)$. Suppose that the points of Π are randomly displaced, in such a way that the displacements of different points are independent, and suppose that the distribution of the displaced position of a point of Π at $X = x$ has a probability density $\rho(x, \cdot)$. Then the displaced points form a Poisson process Π' with rate function λ' given by*

$$\lambda'(y) = \int_{\mathbb{R}^d} \lambda(x)\rho(x, y)\,\mathrm{d}x. \tag{5.25}$$

In particular, if $\lambda(x)$ is a constant λ, and if $\rho(x, y)$ is a function of $y - x$, then $\lambda'(y) = \lambda$ for all y.

Proof The assumptions of the theorem show that the displaced positions Y_X of the points X of Π define a marking of Π. The Marking Theorem shows that

$$\Pi^* = \{(X, Y_X); X \in \Pi\}$$

is a Poisson process on $\mathbb{R}^{2d}$, and the Mapping Theorem then implies that

$$\Pi' = \{Y_X; X \in \Pi\}$$

is a Poisson process on $\mathbb{R}^d$. The mean measure of Π^* has density $\lambda(x)\rho(x, y)$, and that of Π' therefore has density (5.25). (The assumption of densities is only for simplicity of exposition.)

If $\lambda(x) = \lambda$ and $\rho(x, y) = g(y - x)$, then

$$\lambda'(y) = \lambda \int g(y - x)\,\mathrm{d}x = \lambda \int g(x)\,\mathrm{d}x$$

$$= \lambda \int \rho(0, x)\,\mathrm{d}x = \lambda.$$

This persistence property of the homogeneous Poisson process makes it an attractive model in applications involving random displacements. In particular, it has been so applied in a number of biological situations concerned with the spatial distributions of animal or plant populations. Animals displace of their own volition; the displacement of plant populations more often occurs between generations, as seeds are dispersed from the parent plant.

It has however been pointed out by Felsenstein (1975) that enthusiasm for the Poisson process can be carried to excess, even to the point of contradiction. Consider for example an area of the plane in which points represent the positions of a generation of plants. Suppose that those plants die, having distributed seeds to produce the next generation. Then the daughters of a particular plant are scattered around the position of their parent.

If each plant produced at most one daughter, the Displacement Theorem would show that, under very general conditions, a homogeneous Poisson process of parent plants would produce a homogeneous Poisson process of daughters, but this conclusion is false if, as usually happens, some plants produce many offspring.

Specifically, it is often assumed that the number N_x of daughters born to a plant at x has a Poisson distribution with mean $\phi(x)$, and that they are at positions

$$x + D_i \quad (i = 1, 2, \ldots, N_x), \tag{5.26}$$

where the D_i are independent and have probability $g(x, \cdot)$ on $\mathbb{R}^2$.

Under these assumptions, it is easy to see (using in fact the same calculation as was used in Section 2.5 to prove the existence theorem) that the points D_i form a Poisson process with rate $\phi(x)g(x, \cdot)$.

Hence by the Mapping Theorem the daughters of a plant at x form a Poisson process with rate

$$\lambda_x(y) = \phi(x)g(x, y - x). \tag{5.27}$$

Now suppose that the offspring of the different plants behave independently, so that the whole daughter generation consists of the superposition of independent Poisson processes. The Superposition Theorem tells us that the daughter generation is a Poisson process Π' with rate function

$$\lambda'(y) = \sum \phi(x)g(x, y - x), \tag{5.28}$$

where the sum is over all the positions x of the parent plants.

All this, however, is conditional on the positions x of the parents, which is why they have been printed in lower case. If the parents are supposed to lie at the points X of a random set Π (Poisson or not), the analysis shows that the *conditional* distribution of Π' given Π is that of a Poisson process whose rate function

$$\Lambda'(y) = \sum \phi(X)g(X, y - X) \tag{5.29}$$

is itself subject to random variation.

The *unconditional* distribution of Π' will only in trivial cases be Poisson, so that a facile assumption that every generation can be a Poisson process is inconsistent with the assumed biology. We return to this problem in Section 6.2.

5.6 The orbital motorway

Here is another problem about roads, even more over-simplified than that of Section 5.4. A city decides to build a motorway around its built-up area, in a rough circle about the city centre (British readers will think of the M25 around London). Existing radial routes from the centre have grown up haphazardly over the years, and intersect the new road in a random collection of points. Some of these routes are faster than others, and are likely to remain in use with the new road. Some slower roads will fall into disuse for long-distance traffic, if the motorway is fast enough to make it quicker to use faster routes. How many radial routes are likely to remain in use?

Suppose that the radial routes meet the motorway in the points of a Poisson process, whose positions can be described by an angular coordinate θ around the circle. Assume that this process then has a uniform rate λ in $0 \leqslant \theta < 2\pi$. Suppose also that the travel time from the centre to the point θ is a random variable τ_θ, the τ_θ for different θ being independent with a probability density $f(t)$ $(t \geqslant 0)$. Then the τ_θ form a marking of Π, and the points (θ, τ_θ) form a Poisson process Π^* on the plane rectangle $[0, 2\pi) \times (0, \infty)$ with rate

$$\Lambda(\theta, \tau) = \lambda f(\tau). \tag{5.30}$$

Now assume that journeys on the new motorway take place at a constant angular velocity ω so that $2\pi/\omega$ is the total circumnavigation time. Then a route meeting the motorway at θ with $\tau_\theta = \tau$ is viable if there is no other route (θ', τ') with

$$\tau' + \omega^{-1} d(\theta, \theta') < \tau, \tag{5.31}$$

where

$$d(\theta, \theta') = \min(|\theta - \theta'|, 2\pi - |\theta - \theta'|) \tag{5.32}$$

is the angular distance between θ and θ'. Hence the probability $v(\theta, \tau)$ that a point $(\theta, \tau) \in \Pi^*$ is viable is the probability that there are no points $(\theta', \tau') \in \Pi^*$ in the set C defined by the inequality (5.31):

$$v(\theta, \tau) = \exp\left\{ -\iint_C \lambda f(\tau')\, \mathrm{d}\theta'\, \mathrm{d}\tau' \right\},$$

or

$$v(\theta, \tau) = \exp\left\{ -2\lambda \int_0^\tau \min[\omega(\tau - t), \pi] f(t)\, \mathrm{d}t \right\}. \tag{5.33}$$

This is a colouring problem; the points $(\theta, \tau) \in \Pi^*$ are coloured with probability $v(\theta, \tau)$. It is not however possible to conclude that the coloured points form a Poisson process with rate $\Lambda(\theta, \tau)v(\theta, \tau)$, because the colouring events are not independent. Despite this, it is valid to infer that the expected number V of viable routes is

$$V = \int_0^{2\pi} \int_0^{\infty} \Lambda(\theta, \tau)v(\theta, \tau)\, d\theta\, d\tau. \tag{5.34}$$

The easiest way to see this is to condition on the total number N of points of Π^*, which has distribution $\mathscr{P}(2\pi\lambda)$. Conditional upon N, the points

$$(\theta_n, \tau_n), \quad n = 1, 2, \ldots, N$$

of Π^* are independent, with probability density $f(\tau)/2\pi$. The conditional expectation, given N, of the number of coloured points, is thus

$$N \int_0^{2\pi} \int_0^{\infty} \frac{f(\tau)}{2\pi} v(\theta, \tau)\, d\theta\, d\tau,$$

so that

$$V = 2\pi\lambda \int_0^{2\pi} \int_0^{\infty} \frac{f(\tau)}{2\pi} v(\theta, \tau)\, d\theta\, d\tau,$$

which is the same as (5.34).

Substituting for v, (5.34) becomes after simplification

$$V = 2\pi\lambda \int_0^{\infty} f(\tau) \exp\left\{-2\lambda \int_0^{\tau} \min[\omega(\tau - t), \pi] f(t)\, dt\right\}, \tag{5.35}$$

which may in principle be evaluated for any particular density f.

In particular, when λ is large, V may be very much less than the expected total number $2\pi\lambda$ of routes. The precise behaviour of V as $\lambda \to \infty$ depends on the behaviour of f at the left-hand endpoint of its support. If for example f is continuous at $t = 0$ with $f(0) > 0$, the reader will easily check that

$$V \sim 2\pi^{3/2} f(0)^{1/2} \lambda^{1/2} \tag{5.36}$$

as $\lambda \to \infty$.

6

Cox processes

6.1 Definitions and basic properties

In Section 5.5 we met a random set which, while not a Poisson process, could be made into a Poisson process by conditioning. Such processes were introduced by Cox (1955) under the name of *doubly stochastic Poisson processes*, but are now usually given the name of their discoverer.

Thus let the state space S be as in Section 2.1. Let μ be a random non-atomic measure on S, and let Π be a random countable subset of S. Then Π is said to be a *Cox process* associated with μ if the conditional distribution of Π given μ is that of the Poisson process with mean measure μ.

This definition is thus a set of assumptions about the conditional joint distributions of the counts $N(A)$ of Π, given μ. If $A_1, A_2, \ldots, A_n$ are disjoint, then it asserts that

$$\mathbb{P}\{N(A_1) = r_1, N(A_2) = r_2, \ldots, N(A_n) = r_n | \mu\}$$

$$= \prod_{k=1}^{n} \pi_{r_k}(\mu(A_k)). \tag{6.1}$$

The unconditional joint distributions of the $N(A)$ are then obtained by taking expectations.

In most applications the random measure is defined by a density Λ:

$$\mu(A) = \int_A \Lambda(x)\, \mathrm{d}x. \tag{6.2}$$

Thus $\Lambda(x)$ is a real-valued random process on S which must be assumed measurable (jointly in x and the implied ω of the underlying probability space) so that (6.2) is a well-defined random variable. The joint distributions and derived expectations of the $N(A)$ can then be expressed in terms of the joint distributions of the process Λ. For example,

$$\mathbb{E}\{N(A)\} = \mathbb{E}\{\mathbb{E}[N(A)|\mu]\} = \mathbb{E}\int_A \Lambda(x)\, \mathrm{d}x,$$

so that

$$\mathbb{E}\{N(A)\} = \int_A \mathbb{E}\{\Lambda(x)\}\, \mathrm{d}x. \tag{6.3}$$

Sometimes the values of Λ represent some 'real' aspect of the model, like fertility or rate of activity, but in other cases it is a purely mathematical construct. By choosing different processes for Λ many different non-Poisson random sets can be constructed. Thus Cox processes are a very flexible means of producing general random sets which are not too difficult to analyse.

Not every random set is however representable as a Cox process. In particular, the second-order properties of Cox processes satisfy an inequality not fulfilled by general random sets. To see this, compute

$$\begin{aligned}\mathbb{E}\{N(A)^2\} &= \mathbb{E}\{\mathbb{E}[N(A)^2 \mid \mu]\} \\ &= \mathbb{E}\{\mu(A) + \mu(A)^2\} \\ &= \mathbb{E}\{\mu(A)\} + [\mathbb{E}\{\mu(A)\}]^2 + \operatorname{var}\{\mu(A)\},\end{aligned}$$

so that

$$\operatorname{var}\{N(A)\} = \mathbb{E}\{N(A)\} + \operatorname{var}\{\mu(A)\}. \tag{6.4}$$

In particular,

$$\operatorname{var}\{N(A)\} \geqslant \mathbb{E}\{N(A)\}, \tag{6.5}$$

with equality if and only if $\mu(A)$ is a degenerate random variable.

In this sense, all Cox processes are 'over dispersed'. The count $N(A)$ has variance greater than it would have if it were a Poisson variable with the same mean.

A full account of the theory of Cox processes may be found in the book of Grandell (1976), to which the interested reader is referred. We shall here be content with two illustrations, of very different kinds. The first picks up the analysis of Section 5.5, while the second throws some light, in the one-dimensional case, on the extent to which Cox processes are special.

6.2 Cox processes in ecology

In Section 5.5, a random countable set represented the positions of the members of a biological population in a plane habitat. It is likely that there are variations in the fertility or attractiveness of different parts of the habitat, which can be reflected in a non-uniform rate function $\lambda(x)$, large in fertile areas and smaller elsewhere.

If these varying factors can be directly measured, we can postulate that the members of the population form a non-homogeneous Poisson process π, with a rate $\lambda(x)$ which can be considered known. The fact that there may have been random factors at work in determining $\lambda(x)$ can be ignored. On the other hand, if $\lambda(x)$ cannot itself be observed, we must model it too as a random process $\Lambda(x)$. Although Π is, conditional on Λ, a Poisson process, its unconditional structure is that of a Cox process.

Even if there are no differences in fertility, the reproductive mechanism of the population can lead to a Cox, rather than a Poisson process. This is shown by the Felsenstein example of Section 5.5. Take $\phi(X)$ and $g(X, \cdot)$ to be independent of X. Then the daughter generation has been shown to be a Cox process with rate

$$\Lambda(x) = \phi \sum g(x - X) \tag{6.6}$$

depending on the positions X of the parents.

Since this function will always depend on the random variables X, there will always be strict inequality in (6.5) for the daughter generation. Thus the daughter process can never be a Poisson process, if there is any randomness at all in the previous generation. If we follow the population through its successive generations, we see a sequence of non-Poisson Cox processes.

In principle, the distributions of these Cox processes can be computed successively using the random rate functions Λ for the different generations. The analogue of (6.6) for the granddaughter generation is

$$\Lambda'(x) = \phi \sum g(x - X'), \tag{6.7}$$

where the X' are the positions of the daughters. Given Λ, the X' form a Poisson process, so that the distribution of $\Lambda'(x)$, and the joint distributions for different values of x can be calculated from Campbell's Theorem. Thus the successive functions Λ for different generations form a Markov chain whose transition function is in principle known.

The calculations are far too complex to be carried out explicitly, but it is of some interest to consider the first- and second-order moments. By (3.9),

$$\mathbb{E}\{\Lambda'(x)|\Lambda\} = \int_{\mathbb{R}^2} \phi g(x - y)\Lambda(y)\,\mathrm{d}y.$$

Thus

$$\mathbb{E}\{\Lambda'(x)\} = \phi \int_{\mathbb{R}^2} g(x - y)\mathbb{E}\{\Lambda(y)\}\,\mathrm{d}y, \tag{6.8}$$

which shows that, if

$$\mathbb{E}\{\Lambda(x)\} = \lambda, \tag{6.9}$$

a constant, then

$$\mathbb{E}\{\Lambda'(x)\} = \lambda\phi.$$

In particular, if $\phi = 1$, we can have (6.9) in each generation.

Assume therefore that $\phi = 1$ and that (6.9) holds. Then (3.12) shows that

$$\mathbb{E}\{\Lambda'(x)\Lambda'(y)|\Lambda\} = \int_{\mathbb{R}^2} g(x - \xi)\Lambda(\xi)\,\mathrm{d}\xi \int_{\mathbb{R}^2} g(y - \eta)\Lambda(\eta)\,\mathrm{d}\eta$$
$$+ \int_{\mathbb{R}^2} g(x - \xi)g(y - \xi)\Lambda(\xi)\,\mathrm{d}\xi,$$

so that

$$\mathbb{E}\{\Lambda'(x)\Lambda'(y)\} = \int_{\mathbb{R}^2}\int_{\mathbb{R}^2} g(x-\xi)g(y-\eta)\mathbb{E}\{\Lambda(\xi)\Lambda(\eta)\}\,\mathrm{d}\xi\,\mathrm{d}\eta + \lambda\int_{\mathbb{R}^2} g(x-\xi)g(y-\xi)\,\mathrm{d}\xi. \tag{6.10}$$

In particular, if Λ is second-order stationary, with autocovariance function given by

$$\mathbb{E}\{\Lambda(x)\Lambda(y)\} = \lambda^2 + \sigma(y-x), \tag{6.11}$$

then so is Λ' and its autocovariance function σ' satisfies

$$\lambda^2 + \sigma'(y-x) = \int_{\mathbb{R}^2}\int_{\mathbb{R}^2} g(x-\xi)g(y-\eta)\{\lambda^2 + \sigma(\eta-\xi)\}\,\mathrm{d}\xi\,\mathrm{d}\eta + \lambda\int_{\mathbb{R}^2} g(x-\xi)g(y-\xi)\,\mathrm{d}\xi.$$

Writing

$$\gamma(x) = \int_{\mathbb{R}^2} g(y)g(y-x)\,\mathrm{d}y, \tag{6.12}$$

this simplifies to

$$\sigma'(x) = \int_{\mathbb{R}^2} \sigma(x-y)\gamma(y)\,\mathrm{d}y + \lambda\gamma(x), \tag{6.13}$$

which enables the autocovariance functions in successive generations to be computed recursively. It is possible to use the general theory of random walks to show that, if the distribution g has finite variance, then $\sigma(0) = \mathrm{var}\{\Lambda(x)\}$ increases without limit as the generations go by. This is a pecularity of two dimensions (or one); in three or more dimensions we can have $\sigma' = \sigma$ in (6.13).

6.3 The Borel–Tanner distribution

This section is something of a diversion, but it does follow in a sense from Sections 5.5 and 6.2, as well as glimpsing tantalising connections with other aspects of Poisson processes. We have seen that a population which reproduces from generation to generation in a Poisson way displays increasing clumping, although the average density may remain constant. Part of the explanation for this lies in the properties of the successive descendants of a single ancestor.

Thus ignore the spatial distribution, so that the descendants in successive generations are described by a simple (Galton–Watson) branching process in which the family size distribution is $\mathscr{P}(\phi)$. If $\phi \leqslant 1$, the general theory of branching processes tells us that there will only be a finite number of descendants of a given ancestor. Thus even in the critical case $\phi = 1$ the nth generation, for large n, will contain descendants of only a small proportion of the initial generation.

The distribution of the total number of descendants of a single ancestor is easily computed. Consider first a Galton–Watson process (Harris, 1963) with a general family size distribution with generating function $F(z)$. Thus a particular individual has a number N of children, with

$$\mathbb{E}(z^N) = F(z), \tag{6.14}$$

the children have children according to the same distribution, and so on, different family sizes being independent. If the mean family size

$$F'(1) \leqslant 1, \tag{6.15}$$

there are only a finite number of non-empty generations, so that the total progeny M (including the initial ancestor) is finite with probability 1.

Clearly

$$M = 1 + M_1 + M_2 + \cdots + M_N$$

where the M_k are the progeny of the N children of the ancestor. The M_k are independent, with the same distribution as M. Thus $G(z) = \mathbb{E}(z^M)$ satisfies

$$\begin{aligned} G(z) &= \mathbb{E}\{\mathbb{E}(z^M \mid N)\} \\ &= \mathbb{E}\{zG(z)^N\} = zF\{G(z)\}. \end{aligned}$$

This functional equation

$$G = zF(G) \tag{6.16}$$

for G may be solved by Lagrange's expansion (Whittaker and Watson, 1902, Section 7.32). This expresses the solution to (6.16) which is a power series in z as

$$G(z) = \sum_{n=1}^{\infty} \frac{z^n}{n!} \frac{\mathrm{d}^{n-1}}{\mathrm{d}x^{n-1}} [F(x)^n]_{x=0}. \tag{6.17}$$

More generally, for suitable functions f,

$$f(G(z)) = f(0) + \sum_{n=1}^{\infty} \frac{z^n}{n!} \frac{\mathrm{d}^{n-1}}{\mathrm{d}x^{n-1}} [F(x)^n f'(x)]_{x=0},$$

and in particular, for $r \geqslant 1$,

$$G(z)^r = \sum_{n=1}^{\infty} \frac{z^n}{n!} \frac{\mathrm{d}^{n-1}}{\mathrm{d}x^{n-1}} [F(x)^n r x^{r-1}]_{x=0}. \tag{6.18}$$

Since $G(z)^r$ is the generating function of the total progeny $M^{(r)}$ of r sister ancestors, we have

$$\mathbb{P}\{M^{(r)} = n\} = \frac{r}{n!}\frac{\mathrm{d}^{n-1}}{\mathrm{d}x^{n-1}}[F(x)^n x^{r-1}]_{x=0}. \tag{6.19}$$

But $F(x)^n$ is the generating function of the total size S_n of n different families, and (6.19) picks out the coefficient of x^{n-r} with a numerical multiplier. Hence (6.19) implies that

$$\mathbb{P}\{M^{(r)} = n\} = \frac{r}{n}\mathbb{P}\{S_n = n - r\} \quad (n \geqslant r). \tag{6.20}$$

In particular,

$$\mathbb{P}\{M = n\} = \frac{1}{n}\mathbb{P}\{S_n = n - 1\}. \tag{6.21}$$

These results have a number of different applications, for instance in queueing theory where $M^{(r)}$ is the number of customers in a busy period when r customers are initially present. The equations have a combinatorial interpretation close to the ballot theorems mentioned in Section 4.3.

Now specialise to the case studied in the last section, in which the family size distribution is $\mathscr{P}(\phi)$ for $\phi \leqslant 1$. Then S_n has distribution $\mathscr{P}(n\phi)$, so that (6.20) and (6.21) become

$$\mathbb{P}\{M^{(r)} = n\} = \frac{r}{n}\pi_{n-r}(n\phi), \tag{6.22}$$

$$\mathbb{P}\{M = n\} = \frac{1}{n}\pi_{n-1}(n\phi). \tag{6.23}$$

Even the fact that these are probability distributions is analytically non-trivial. Thus (6.22) implies that, for all $r \geqslant 1$ and $t \leqslant 1$,

$$\sum_{n=r}^{\infty}\frac{\pi_{n-r}(nt)}{n} = \frac{1}{r}. \tag{6.24}$$

The special case $r = 1$, written out in full, is

$$\sum_{n=1}^{\infty}\frac{n^{n-1}t^{n-1}\,\mathrm{e}^{-nt}}{n!} = 1,$$

or

$$\sum_{n=1}^{\infty}\frac{n^{n-1}}{n!}(t\,\mathrm{e}^{-t})^n = t \quad (t \leqslant 1). \tag{6.25}$$

In other words, the function

$$\psi(x) = \sum_{n=1}^{\infty} \frac{n^{n-1}}{n!} x^n, \tag{6.26}$$

defined by this convergent series when $|x| \leqslant e^{-1}$ (Stirling's formula), satisfies

$$\psi(t\,e^{-t}) = t \tag{6.27}$$

when $0 \leqslant t \leqslant 1$. It therefore solves the functional equation

$$t\,e^{-t} = s \tag{6.28}$$

in this region, by $t = \psi(s)$.

This of course is the same answer as we would have obtained by a direct use of Lagrange's expansion, so that the identity (6.24) actually contains this special case of Lagrange's result.

The distribution (6.22) was derived by Borel and generalised to (6.23) by Tanner (1953); it is therefore known in the literature of the theory of queues as the *Borel–Tanner distribution.*

6.4 Cox processes and renewal processes

We now return to Cox processes, and confine attention to the one-dimensional case $S = \mathbb{R}$. This is special since the general non-homogeneous process can be constructed from a homogeneous process by a monotone transformation.

Specifically, let λ be a positive function such that

$$\int_{-\infty}^{0} \lambda(x)\,dx = \int_{0}^{\infty} \lambda(x)\,dx = \infty. \tag{6.29}$$

Let $\psi\colon \mathbb{R} \to \mathbb{R}$ be the unique right-continuous function with

$$\int_{0}^{\psi(y)} \lambda(x)\,dx = y. \tag{6.30}$$

Let Π be a Poisson process of constant 1 on $\mathbb{R}$. Then (Section 4.5) the points $\psi(X)$ $(X \in \Pi)$ form a Poisson process with rate $\lambda(x)$.

To construct a Cox process we replace λ by a (measurable) random process Λ independent of Π. Then (6.30) defines a right-continuous increasing process Ψ, and

$$\{\Psi(X);\, X \in \Pi\} \tag{6.31}$$

is a Cox process corresponding to Λ.

This construction can be looked at in two ways. It uses Ψ to *transform* the homogeneous Poisson process Π. Alternatively, it *samples* the process Ψ at Poisson instants $X \in \Pi$. The second viewpoint is important because Poisson sampling loses no information about joint distributions. If we know the joint distributions of the points of (6.31), we can deduce those of the process Ψ and so work back to those of Λ.

This is by no means obvious, and we refer the interested reader to Kingman (1964). There the result is applied to the question of which renewal processes can be represented as Cox processes. By a renewal process is meant a random subset

$$0 < X_1 < X_2 < \cdots \tag{6.32}$$

of $(0, \infty)$ such that the variables

$$X_1, X_2 - X_1, X_3 - X_2, \ldots \tag{6.33}$$

are independent, all but perhaps the first having a common 'lifetime' distribution. It turns out that, for (6.32) to be a Cox process, the lifetime distribution must be of a very special form. It must have a probability density which can be written

$$\lambda p(t), \tag{6.34}$$

where $\lambda > 0$ and p is a 'p-function' (Kingman 1972), a function of the form

$$p(t) = \mathbb{P}\{\xi(t) = 0 \,|\, \xi(0) = 0\} \tag{6.35}$$

for some Markov process ξ with $p(0+) = 1$. The corresponding rate process is of the form

$$\begin{aligned} \Lambda(t) &= \lambda \qquad \text{if } \xi(t) = 0, \\ &= 0 \qquad \text{otherwise}. \end{aligned} \tag{6.36}$$

For our purposes this result emphasises the special character of Cox processes, and guards against the complacent feeling that 'most' interesting random countable sets are, if not Poisson, at least members of the Cox family.

7

Stochastic geometry

7.1 Poisson processes of geometrical objects

So far we have considered random countable subsets of spaces S which have always been subsets of $\mathbb{R}^d$ for some dimension d. The theory is, however, much more general, and applies to random structures which look quite different from the irregular array of isolated points depicted in Fig. 1.1.

Suppose for instance that one looks through a microscope at a piece of paper. The view might roughly resemble Fig. 7.1, where the lines represent fibres of the material. This could be modelled by representing the fibres as infinite straight lines, or in a more complex way allowing for thickness, finite length or departure from straightness. The lines could be regarded as lying in a plane, or in three dimensions if the thickness of the material justified it.

Applications like this show a need for models for random arrays of lines in $\mathbb{R}^2$ and $\mathbb{R}^3$. Similarly one could consider arrays of planes in $\mathbb{R}^2$ or of higher dimensional affine subspaces in $\mathbb{R}^d$ for larger d. Nor are flat subsets the only ones of possible interest; the wet pavement problem of Section 5.2 can be regarded as a random array of discs in $\mathbb{R}^2$.

Such random collections of geometrical objects are the subject of *stochastic geometry*, a term roughly equivalent to *geometrical probability* or *integral geometry* although these different names signal differences in emphasis.

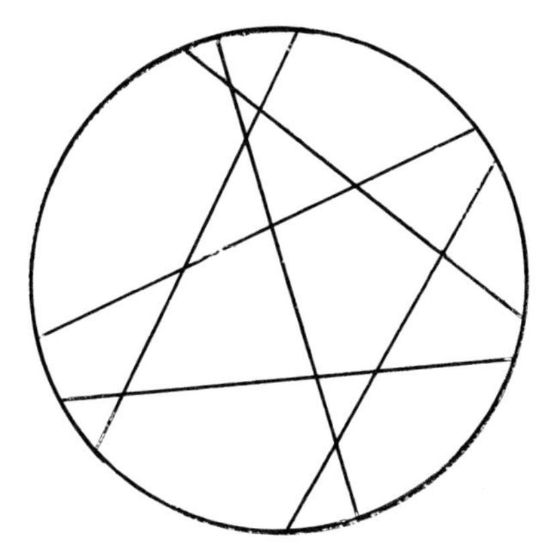

Fig. 7.1 Random lines in a plane.

The key to understanding them is to realise that, in most cases of importance, the geometrical objects in question form a set S of sufficiently simple structure that the random collection can be studied as a random subset of S. The whole apparatus of Poisson theory can then at once be brought to bear.

Thus suppose for example that Fig. 7.1 is to be modelled as a random collection of infinite (undirected) straight lines in $\mathbb{R}^2$. The set S of such lines is a two-dimensional manifold (a space which is locally equivalent to $\mathbb{R}^2$), since two parameters suffice to determine a line in the plane. Thus each line is a 'point' in S, and a random collection of lines is a random subset of S.

This situation will be examined in more detail in the next section, where it will be seen that a subset of S which looks like Fig. 7.1, with only finitely many lines in any bounded field of view, is one which is finite on compact subsets of S. One way of generating such random subsets is by means of the Existence Theorem, starting from a mean measure which is non-atomic and finite on compact subsets.

To every such measure μ on S corresponds a *Poisson line process*, which is simply a Poisson process on the set S of lines. Exactly similarly, we could define a Poisson line process in $\mathbb{R}^3$ by taking S to be the set of lines in $\mathbb{R}^3$, a four-dimensional manifold.

Again, we might be interested in a random array of planes in $\mathbb{R}^3$, in which case S is three-dimensional. More generally, a random array of d-dimensional affine subspaces in $\mathbb{R}^n$ could be represented as a Poisson process on a manifold of dimension $(d + 1)(n - d)$.

The geometrical objects need not be flat. The discs in $\mathbb{R}^2$ (the raindrops on the pavement) need three parameters to define them; they form a set which is not quite a 3-manifold because of a boundary at zero radius, but this is not a serious obstacle because integration ignores such anomalies. Arrays of curves and surfaces in $\mathbb{R}^2$ and $\mathbb{R}^3$ can be dealt with in a similar way if they can be described by a finite number of parameters.

This is a book about Poisson processes rather than stochastic geometry, and we shall not therefore do more than illustrate what has become a rich theory. The next two sections deal with line processes in the plane, and we return briefly to more general problems in the final section. The reader who wishes to explore further may find useful the bibliographical comments at the end of the chapter.

7.2 Line processes

In this section and the next, S denotes the set of infinite, undirected straight lines in $\mathbb{R}^2$. A random array of lines is then just a random subset of S and will be called a *line process*. If the random subset is a Poisson process, it will be called a *Poisson line process*.

A Poisson line process has a mean measure μ which is a measure on S, and calculations involving this measure need a coordinate system on S. There are various ways of setting up such a system, but a useful one for our purposes is as follows.

If l is any line in $\mathbb{R}^2$, there is a unique line $l^{\perp}$ through the origin O perpendicular to l. Let L be the intersection of l and $l^{\perp}$, and let p be the distance OL, taken as positive if L lies in the upper half-plane $\{(x, y); y > 0\}$ or on the positive x-axis $\{(x, 0); x > 0\}$ and negative otherwise. Let θ be the angle that $l^{\perp}$ makes with the axis, taken such that $0 \leqslant \theta < \pi$.

The function $l \mapsto (p, \theta)$ maps S in a one-to-one way onto the strip

$$\tilde{S} = \{(p, \theta); -\infty < p < \infty, 0 \leqslant \theta < \pi\} \tag{7.1}$$

(the reader should draw a figure). Thus any measure on S induces a measure on $\tilde{S}$ and vice versa; by a convenient abuse of language we do not distinguish these two measures.

If

$$D_r = \{(x, y); x^2 + y^2 \leqslant r^2\}$$

is the disc with centre O and radius r, then the line l meets D_r if and only if $|p| \leqslant r$. From this it follows easily that a subset A of S is a set of lines of which only finitely many meet any bounded set if and only if A maps into a subset $\tilde{A}$ of $\tilde{S}$ which is locally finite (has finite intersection with every bounded set).

In particular, if μ is a non-atomic measure on $\tilde{S}$ which is finite on bounded sets, the Existence Theorem shows that there is a Poisson line process with mean measure μ and that, with probability 1, only finitely many of the random lines meet any bounded subset of $\mathbb{R}^2$.

One obvious example is Lebesgue measure (area) on $\tilde{S}$

$$\mu_1(\tilde{A}) = \int_{\tilde{A}} \mathrm{d}p \, \mathrm{d}\theta, \tag{7.2}$$

or more generally

$$\mu_\lambda = \lambda \mu_1. \tag{7.3}$$

A Poisson line process with mean measure μ_λ is called a *uniform* Poisson line process with rate λ.

The word 'uniform' is justified by a very important invariance property of μ_λ which implies a stochastic invariance of the line process itself. The Euclidean motions of the plane, the mappings which preserve distance, can be built up from translations, rotations about O and reflections. Each of these maps lines into lines, so inducing mappings $S \to S$ and so $\tilde{S} \to \tilde{S}$. Each of these leaves the measure μ_λ invariant.

To see this, consider first a translation which sends O to the point (ξ, η). It is easily checked that the line with coordinates (p, θ) maps into the line with coordinates

$$(p + \xi \cos \theta + \eta \sin \theta, \theta). \tag{7.4}$$

This is a non-uniform shear of $\tilde{S}$ which is clearly area-preserving.

A rotation about O through an angle α sends (p, θ) into the line

$$((-1)^k p, \theta + \alpha - k\pi) \tag{7.5}$$

where the integer k is such that $0 \leqslant \theta + \alpha - k\pi < \pi$.

It is a little more difficult to see that this preserves area, since k depends on θ. However, for given α, k takes on at most two values, so that any set can be split into two sets each of which is rotated by the transformation. Thus (7.5) also preserves area, and thus μ_λ.

We leave the reader to check that μ_λ is also preserved by reflections, so that it is invariant under all Euclidean motions. If therefore a line process Π' is derived from a uniform line process Π by such a motion, then by the Mapping Theorem, Π' is a uniform Poisson line process with the same rate.

It is not difficult to show that the μ_λ for $\lambda \geqslant 0$ are the only measures with this invariance property, a point to which we return in Section 7.4.

Some calculations about uniform Poisson processes are almost trivial consequences of the definition. Suppose for instance that we are interested in the number N of lines of Π which meet D_r. This is the number of 'points' which fall in the corresponding subset A of S, which has the Poisson distribution with mean $\mu_\lambda(A)$. Since $\tilde{A} = \{(p, \theta); |p| \leqslant r\}$ we have $\mu_\lambda(A) = 2\lambda r$, so that N has distribution $\mathscr{P}(2\lambda r)$.

More generally, the number N which meets a bounded set $B \subset \mathbb{R}^2$ has distribution $\mathscr{P}(\mu)$, where

$$\mu = \lambda \iint \mathrm{d}p \, \mathrm{d}\theta, \tag{7.6}$$

the integral being taken over all (p, θ) for which the corresponding line meets B. The argument extends; the number of lines which meet each of two bounded sets for instance has distribution $\mathscr{P}(\mu)$, where μ is given by (7.6) with the appropriate range of integration. Any difficulty lies in evaluating (7.6), and is therefore geometric rather than probabilistic.

Much more difficult are problems of the geometry of two or more lines of the process. For instance, the points of intersection of all the pairs of lines of Π certainly do not form a Poisson process (indeed, a Poisson process in $\mathbb{R}^2$ has, with probability 1, no collinear triples). The stochastic structure of this point process is very complicated, but there are even more subtle geometrical questions that can be asked of this random array of lines. For instance, the lines of the uniform Poisson line process Π divide the plane into disjoint polygons; what proportion of these are triangles, quadrilaterals, pentagons and so on? There are answers to these questions, notably in the work of Miles (1969, 1971). To tantalise the reader with just one example, the mean number of sides of these random polygons is 4.

7.3 Cox line processes

For line processes as for point processes, there are more general models than simple Poisson processes. One very useful class is that of Cox processes, defined exactly as in Chapter 6. Thus Π is a Cox line process if there is a random measure μ on the set S of lines such that, conditional on μ, Π is a Poisson process with mean measure μ.

In an important sequence of papers (collected in Harding and Kendall, 1974) shortly before his early death, Rollo Davidson examined the properties of those Poisson and Cox line processes which are invariant under all translations of the plane. He encountered surprising difficulty in constructing invariant processes which were *not* Cox processes, and those he did construct were either pathological or degenerate in the sense of having, with positive probability, pairs of parallel lines. It is clear that Cox processes come much nearer to exhausting all (invariant) line processes than is the case for point processes.

Davidson was led to conjecture that any invariant line process with finite mean square (of counts on bounded sets) and with no parallel lines is necessarily a Cox process. The conjecture is false, but it took a very ingenious construction by Kallenberg (1977) to show this, and Kallenberg's construction is itself degenerate in a more subtle sense.

There remains therefore something of a mystery, which seems to be connected in a curious way with concepts of transitivity in the various relevant groups of transformations. The reader might like to read Kallenberg's paper and the references cited in it.

7.4 More general geometrical objects

It will be clear from the special case of lines in the plane that a general theory of Poisson processes of geometrical objects would need to involve a number of distinct steps. First we must describe the set S of objects of interest. If the objects can be smoothly described by a finite set of parameters, S will be a manifold of some dimension D, not necessarily the same as that of the space in which the objects sit. The parametrisation may have discontinuities arising from the topology of S and the manifold may have a boundary, but these will only complicate probabilistic calculations if they are of positive measure.

A Poisson process of objects of S is then defined simply as a random subset Π of S with the properties of independence and Poisson distribution set out in Section 2.1. This process is described completely by its mean measure μ, so that we need a convenient way of handling measures μ on the manifold S. Typical probabilities will be derived from Poisson distributions whose means $\mu(A)$ need to be calculated for sets A corresponding to some geometrical property.

Thus we need to be able to translate interesting geometrical properties of the objects into the chosen parametrisation of S, and then carry out the necessary integration. Alternatively, it may sometimes be possible to avoid explicit integration, for instance by symmetry arguments, and thus to circumvent the parametrisation.

These operations are the subject matter of *integral geometry*, or under another name *geometrical probability*. A number of ingenious techniques have been devised, notably by Morgan Crofton, to avoid complicated multiple integration, and these have led to surprising and elegant results.

A central concept is that of invariance under a group of transformations of the underlying space. These transformations induce transformations of the manifold S and the latter form a group G acting on S. In typical cases, G is a topological group acting transitively on S (so that every point of S is mapped onto any other point by some element of G), which means that S is a *homogeneous space* under G.

In these circumstances, there are very general theorems that guarantee the existence of a *relatively invariant measure* on S, a measure μ that is transformed by any element g of G into a multiple $c(g)\mu$ of μ. This measure μ is uniquely determined up to constant multiples, and is therefore the natural measure on S in the presence of the group G.

It is not in general true that this measure is *invariant* (i.e. that $c(g) = 1$ for all g). For example, if $S = \mathbb{R}^d$ and G consists of all non-singular affine transformations, then μ is Lebesgue measure and $c(g)$ is the determinant of g^{-1}. In this situation there can be no non-empty Poisson process invariant under G. This is a sign that G is too large, and the subgroup $\{g \in G; c(g) = 1\}$ may be a more natural group of transformations to use.

If μ is invariant, then a Poisson process whose mean measure is a multiple of μ is invariant under G and would then be described as a uniform Poisson process of the objects of S. The Croftonesque techniques for computing $\mu(A)$ are especially powerful when μ is invariant under a suitable group.

The reader who wishes to pursue these matters will find the general theory in Nachbin (1965) or Santaló (1976), the latter with fuller arrays of examples. A more accessible account of the simpler cases will be found in Kendall and Moran (1963). It is well worth reading Crofton's article on 'Probability' in the ninth edition of the *Encyclopaedia Britannica*.

8

Completely random measures

8.1 The canonical representation

If Π is a Poisson process on S, the count function

$$N(A) = \#\{\Pi \cap A\} \tag{8.1}$$

is a random measure on S, since for disjoint $A_j \subset S$,

$$N\left(\bigcup_{j=1}^{\infty} A_j\right) = \sum_{j=1}^{\infty} N(A_j). \tag{8.2}$$

It has also the very special property that the random summands on the right of (8.2) are independent random variables.

This suggests the concept of a *completely random measure* on S. This is a random function Φ from the collection of measurable subsets of S into $[0, \infty]$ such that, for any collection of disjoint measurable sets $A_1, A_2, A_3, \ldots$, the random variables $\Phi(A_j)$ are independent, and

$$\Phi\left(\bigcup_{j=1}^{\infty} A_j\right) = \sum_{j=1}^{\infty} \Phi(A_j). \tag{8.3}$$

To avoid pathologies, we shall assume the standard condition that the diagonal (2.2) is measurable, and that $\Phi(\varnothing) = 0$.

The count function of a Poisson process is a completely random measure, but with the additional property that its values are integers. Without this additional condition, a much richer variety of completely random measures can be constructed, but it turns out they all come, in a sense to be made precise, from Poisson processes.

Note first that, for any completely random measure Φ, we can repeat the argument following (2.4) in Section 2.1. This shows that the joint distributions of random variables

$$\Phi(A_j) \quad (j = 1, 2, \ldots, n),$$

where the sets A_j are not necessarily disjoint, are determined once the distribution of $\Phi(A)$ is known for all $A \subseteq S$.

A convenient way of describing this distribution is by means of the function

$$\lambda_t(A) = -\log \mathbb{E}\{e^{-t\Phi(A)}\} \tag{8.4}$$

for $t > 0$. Clearly $0 \leqslant \lambda_t(A) \leqslant \infty$ and (for some and then for all $t > 0$)

$$\lambda_t(A) = 0 \quad \text{if and only if } \mathbb{P}\{\Phi(A) = 0\} = 1, \tag{8.5}$$

$$\lambda_t(A) = \infty \quad \text{if and only if } \mathbb{P}\{\Phi(A) = \infty\} = 1. \tag{8.6}$$

If the A_j are disjoint, (8.3) and independence show that

$$\begin{aligned}\exp\left\{-\lambda_t\left(\bigcup_{j=1}^{\infty} A_j\right)\right\} &= \mathbb{E}\left\{\exp\left[-t\sum_{j=1}^{\infty}\Phi(A_j)\right]\right\} \\ &= \mathbb{E}\left\{\prod_{j=1}^{\infty}\exp[-t\Phi(A_j)]\right\} \\ &= \prod_{j=1}^{\infty}\exp\{-\lambda_t(A_j)\},\end{aligned}$$

so that

$$\lambda_t\left(\bigcup_{j=1}^{\infty} A_j\right) = \sum_{j=1}^{\infty}\lambda_t(A_j). \tag{8.7}$$

Since $\lambda_t(\emptyset) = 0$, λ_t is a measure on S for each $t > 0$. From (8.5) the λ_t are mutually absolutely continuous, and (8.6) shows that the λ_t are finite or infinite together.

We now impose a further restriction, that the measure λ_t (for some, and then for all, $t > 0$) be σ-finite. In view of (8.6) this requires that there be a countable dissection

$$S = \bigcup_{j=1}^{\infty} S_j$$

of S such that $\Phi(S_j)$ is finite with positive probability. This is not the same as saying that Φ is almost surely σ-finite, and we shall say that Φ is Σ-finite when it holds.

If Φ is Σ-finite, the λ_t have the same atoms, which are called the *fixed atoms* of Φ. Thus x is a fixed atom of Φ if and only if

$$\mathbb{P}\{\Phi(\{x\}) > 0\} > 0. \tag{8.8}$$

The fixed atoms form at most a countable subset $\{a_1, a_2, \ldots\}$; let S_0 be its complement in S. Then it is easy to check that the restriction Φ_0 of Φ to S_0 is a completely random measure without fixed atoms, and is independent of the independent random variables

$$\Phi(\{a_j\}) \quad (j = 1, 2, \ldots).$$

Thus the fixed atoms can be removed by simple surgery, and there is no loss of generality in supposing a Σ-finite completely random measure to have no fixed atoms, so that each λ_t is non-atomic. Let $A \subset S$ have $l = \lambda_1(A) < \infty$. Then a well-known result (a special case of Lyapunov's theorem, see Halmos 1950, page 174) ensures the existence of a dissection

$$A_{nj} \quad (j = 1, 2, \ldots, n)$$

of A such that $\lambda_1(A_{nj}) = l/n$. Thus

$$\mathbb{E}\{e^{-\Phi(A_{nj})}\} = e^{-l/n},$$

whence, for any $c > 0$,

$$\mathbb{P}\{\Phi(A_{nj}) \geqslant c\} \leqslant \frac{1 - e^{-l/n}}{1 - e^{-c}}$$

which tends to zero uniformly in j as $n \to \infty$. Since

$$\Phi(A) = \sum_{j=1}^{n} \Phi(A_{nj})$$

and the summands are independent, the random variable $\Phi(A)$ is infinitely divisible (Doob 1953).

The Lévy–Khinchin representation for infinitely divisible random variables specialises, for positive variables like $\Phi(A)$, to the form

$$\mathbb{E}\{e^{-t\Phi(A)}\} = \exp\left\{-\beta(A)t + \int_{(0,\infty]} (1 - e^{-tz})\gamma(A, dz)\right\}, \tag{8.9}$$

for $t > 0$, where $\beta \geqslant 0$ and $\gamma(A, \cdot)$ is a measure on $(0, \infty]$ which makes the integral in (8.9) converge. Hence

$$\lambda_t(A) = \beta(A)t + \int_{(0,\infty]} (1 - e^{-tz})\gamma(A, dz), \tag{8.10}$$

and (8.10) determines $\beta(A)$ and $\gamma(A, \cdot)$ uniquely in terms of $\lambda_t(A)$ $(t > 0)$.

Substituting (8.10) into (8.7) and using the uniqueness, we have

$$\beta\left(\bigcup_{j=1}^{\infty} A_j\right) = \sum_{j=1}^{\infty} \beta(A_j) \tag{8.11}$$

and

$$\gamma\left(\bigcup_{j=1}^{\infty} A_j, \cdot\right) = \sum_{j=1}^{\infty} \gamma(A_j, \cdot). \tag{8.12}$$

We therefore have a representation for the distributions of the most general Σ-finite completely random measure without fixed atoms; they are given by (8.9) and (8.10), where β is a measure on S and γ is a set function of two arguments, such that

(i) for $A \subseteq S$, $\gamma(A, \cdot)$ is a measure on $(0, \infty]$,

(ii) for $B \subseteq (0, \infty]$, $\gamma(\cdot, B)$ is a measure on S.

Both β and γ must be non-atomic on S, and γ must make the integral in (8.10) converge, for some and then for all $t > 0$, at least on the sets S_j.

To complete the picture, we should have a converse result, showing how to construct a completely random measure from β and γ. This will be done in the next section, which will demonstrate the close connection with Poisson processes.

8.2 Construction from Poisson processes

There is an obvious kinship between equation (8.9) and Campbell's Theorem, especially in the version of Section 5.3, and this is the key to the construction of a completely random measure from β and γ satisfying the conditions of the last section. Forget for the moment about β and concentrate on γ, a function of two arguments $A \subseteq S$ and $B \subseteq (0, \infty]$ which is a measure in each argument when the other is held fixed. Suppose first that the measure defined on S by

$$\mu(A) = \gamma(A, (0, \infty]) \tag{8.13}$$

is σ-finite.

For $z > 0$, the measure

$$\mu_z(A) = \gamma(A, (0, z]) \tag{8.14}$$

is absolutely continuous with respect to μ and so has a Radon–Nikodym derivative $F(x, z)$:

$$\mu_z(A) = \int_A F(x, z)\mu(\mathrm{d}x). \tag{8.15}$$

It is important to recall that (8.15) only determines $F(\cdot, z)$ almost everywhere for each z; a standard argument (Doob 1953) shows that a version can be taken for each rational z and then extended to all z by right-continuity so that, for each x, $F(x, z)$ is the distribution function of a random variable with values in $(0, \infty]$.

We can now define a measure on the product space

$$S^* = S \times (0, \infty] \tag{8.16}$$

by

$$\mu^*(A^*) = \iint_{A^*} \mathrm{d}F(x, z)\mu(\mathrm{d}x), \tag{8.17}$$

where $\mathrm{d}F$ denotes the Stieltjes measure with respect to z. If $A^* = A \times B$, this becomes

$$\mu^*(A \times B) = \int_A \int_B \mathrm{d}F(x, z)\mu(\mathrm{d}x)$$

$$= \int_B \mathrm{d}\gamma(A, (0, z]) = \gamma(A, B).$$

In general, (8.13) will not be σ-finite, but γ must satisfy

$$\int (1 - \mathrm{e}^{-z})\gamma(S_j, \mathrm{d}z) < \infty,$$

so that $\gamma(S_j, (\varepsilon, \infty]) < \infty$ for any $\varepsilon > 0$. It follows that

$$\mu^{(k)}(A) = \gamma\left(A, \left(\frac{1}{k+1}, \frac{1}{k}\right]\right)$$

is a σ-finite measure, and

$$\mu = \sum_{k=0}^{\infty} \mu^{(k)}.$$

Applying the argument to $\mu^{(k)}$ and then summing over k we see that, quite generally, there is a measure μ^* on S^* such that

$$\mu^*(A \times B) = \gamma(A, B), \tag{8.18}$$

for $A \subseteq S$, $B \subseteq (0, \infty]$.

Now let Π^* be a Poisson process on S^* with mean measure μ^*, and define

$$\Psi(A) = \sum \{z; (x, z) \in \Pi^*, x \in A\}. \tag{8.19}$$

Thus Ψ is a purely atomic measure on S whose atoms correspond to the points of Π^*: if (x, z) is a point of Π^* then Ψ has an atom of weight z at x. Clearly the values of Ψ on disjoint sets are independent, so that Ψ is a completely random measure. The distributions of Ψ are easily computed by Campbell's Theorem; for $t > 0$

$$\mathbb{E}\{\mathrm{e}^{-t\Psi(A)}\} = \exp\left\{-\int_A \int_{(0,\infty]} (1 - \mathrm{e}^{-tz})\mu^*(\mathrm{d}x\,\mathrm{d}z)\right\}$$

$$= \exp\left\{-\int_{(0.\infty]} (1 - \mathrm{e}^{-tz})\gamma(A, \mathrm{d}z)\right\}. \tag{8.20}$$

Comparing (8.20) with (8.9), we see that $\Phi(A)$ has the same distribution as $\beta(A) + \Psi(A)$. In other words, this construction produces the most general Σ-finite completely random measure without fixed atoms, in the form of the sum of a deterministic measure β and a purely atomic measure Ψ derived as in (8.19) from a Poisson process on the product space S^*. If we add back the fixed atoms, we see that every Σ-finite completely random measure has the same stochastic structure as the sum of a deterministic measure and a purely atomic measure.

The construction of Ψ almost shows it to be derived from a marked Poisson process. Not quite, however, because the projection μ of μ^* onto S need not be σ-finite. We shall meet in the next chapter a very important completely random measure which is Σ-finite, but has $\mu(A) = \infty$ for all A of positive measure.

8.3 The Blackwell argument

In the last section we uncovered a very remarkable property of (Σ-finite) completely random measures. Apart from a deterministic component β, such a measure has the same distribution as a purely atomic measure Ψ. What we have not proved is that if $\beta = 0$, then Φ is purely atomic with probability 1. This is a much deeper matter, which was addressed by Blackwell (1973) for a special case (the gamma process to be described in Chapter 9). His argument may be extended to a rather large class of completely random measures.

Blackwell assumes (cf. Section 3.4) that S is a Borel subset of some complete separable metric space. This assumption is stronger than that of Section 2.1, but is satisfied by all the spaces we have encountered, for instance the geometrical manifolds of Chapter 7. Its significance is that it implies the existence of a sequence $B_1, B_2, B_3, \ldots$ of measurable subsets of S which is *separating* in the sense that, for any distinct $x_1, x_2 \in S$, there is a value of n such that B_n contains one but not both of x_1, x_2.

Thus each $x \in S$ can be identified with the sequence $(\xi_1, \xi_2, \ldots)$, where

$$\xi_n = \xi_n(x) = 1 \text{ if } x \in B_n, \quad = 0 \text{ otherwise.}$$

More formally, the function

$$\xi \colon x \mapsto (\xi_1(x), \xi_2(x), \ldots) \tag{8.21}$$

is an injection of S into the space $\{0, 1\}^\infty$ of infinite sequences of 0's and 1's. We shall not distinguish between x and $\xi(x)$, and we shall write

$$x_n = (\xi_1(x), \xi_2(x), \ldots, \xi_n(x)) \tag{8.22}$$

for the beginning of the ξ-sequence. For any finite sequence

$$\sigma = (\varepsilon_1, \varepsilon_2, \ldots, \varepsilon_n) \tag{8.23}$$

of 0's and 1's, denote by $\langle \sigma \rangle$ the subset

$$\langle \sigma \rangle = \{x \in S;\, x_n = \sigma\}. \tag{8.24}$$

Now let Φ be a random measure on S with

$$\mathbb{P}\{\Phi(S) < \infty\} = 1. \tag{8.25}$$

Blackwell says that a point $x \in S$ *conforms to* Φ *at stage* $n + 1$ if

$$\xi_{n+1}(x) = 1 \textit{ if and only if } \Phi\langle x_n 1\rangle \geqslant \Phi\langle x_n 0\rangle. \tag{8.26}$$

(Here $x_n 1$ is the sequence of length $n + 1$ obtained by adding 1 to the end of the sequence x_n, and parentheses are omitted after Φ for clarity.) He says that x *ultimately conforms to* Φ if it conforms at all but a finite number of stages.

For any Φ, the set of points ultimately conforming to Φ is clearly a countable random set A. If it can be proved that

$$\mathbb{P}\{\Phi(A) = \Phi(S)\} = 1, \tag{8.27}$$

then Φ must be purely atomic, its atoms all lying in A.

To prove (8.27), let X be a random variable with values in S which, conditional on Φ, has distribution $\Phi(\cdot)/\Phi(S)$. Then (8.27) is equivalent to

$$\mathbb{P}\{X \in A\} = 1, \tag{8.28}$$

that X almost certainly ultimately conforms to Φ. The probability that X fails to conform at stage $n + 1$, conditional on X_n and on Φ, is

$$w_n = \min\left\{\frac{\Phi\langle x_n 1\rangle}{\Phi\langle x_n\rangle}, \frac{\Phi\langle x_n 0\rangle}{\Phi\langle x_n\rangle}\right\}. \tag{8.29}$$

The unconditional probability is $\mathbb{E}\{w_n\}$ and hence (8.28) follows from the Borel–Cantelli lemma if

$$\sum_{n=0}^{\infty} \mathbb{E}\{w_n\} < \infty. \tag{8.30}$$

But

$$\begin{aligned}\mathbb{E}\{w_n\} &= \mathbb{E}\left\{\sum_{\sigma} \frac{\Phi\langle\sigma\rangle}{\Phi(S)} \min\left[\frac{\Phi\langle\sigma 1\rangle}{\Phi\langle\sigma\rangle}, \frac{\Phi\langle\sigma 0\rangle}{\Phi\langle\sigma\rangle}\right]\right\}\\ &= \mathbb{E}\left\{\sum_{\sigma} \min[\Phi\langle\sigma 1\rangle, \Phi\langle\sigma 0\rangle]/\Phi(S)\right\},\end{aligned} \tag{8.31}$$

where the sum extends over all sequences (8.23) of length n.

In the special case considered by Blackwell, he is able to bound (8.31) by direct computation, and to deduce (8.30). For more general Φ it is convenient to proceed slightly differently. Suppose that Φ is a completely random measure for which the deterministic component β is zero, and strengthen (8.25) to

$$\mathbb{E}\{\Phi(S) < \infty\}. \tag{8.32}$$

The Borel–Cantelli lemma applies equally well (unless $\Phi(S) = 0$, in which case the result is trivial) if we can prove that

$$\sum_{n=\infty}^{\infty} \mathbb{E}\{w_n \Phi(S)\} < \infty, \tag{8.33}$$

and (8.31) implies that

$$\mathbb{E}\{w_n \Phi(S)\} = \mathbb{E}\left\{\sum_{\sigma} \min[\Phi\langle\sigma 1\rangle, \Phi\langle\sigma 0\rangle]\right\}. \tag{8.34}$$

The construction of Section 8.2 produces a purely atomic measure Ψ from a Poisson process Π^* on S^*, such that Φ and Ψ have the same joint distributions. By carrying out the construction twice, we can produce two independent such measures Ψ_1, Ψ_2 having the same distributions as Φ. Since $\Phi\langle\sigma 1\rangle$ and $\Phi\langle\sigma 0\rangle$ are independent,

$$\begin{aligned}\mathbb{E}\{\min[\Phi\langle\sigma 1\rangle, \Phi\langle\sigma 0\rangle]\} &= \mathbb{E}\{\min[\Psi_1\langle\sigma 1\rangle, \Psi_2\langle\sigma 0\rangle]\} \\ &\leqslant \mathbb{E}\{\min[\Psi_1\langle\sigma\rangle, \Psi_2\langle\sigma\rangle]\}.\end{aligned}$$

Hence

$$\mathbb{E}\{w_n \Phi(S)\} \leqslant \mathbb{E}\left\{\sum_{\sigma} \min[\Psi_1\langle\sigma\rangle, \Psi_2\langle\sigma\rangle]\right\}. \tag{8.35}$$

The right-hand side of this equation can be regarded in the following way. Suppose that we want to compute

$$\Upsilon = \Psi_1 \wedge \Psi_2, \tag{8.36}$$

the largest measure on S which lies below Ψ_1 and Ψ_2. We can do this by looking at the restrictions of Ψ_1 and Ψ_2 to the finite σ-algebra $\mathscr{B}_n$ generated by $B_1, B_2, \ldots, B_n$, and taking the lesser of Ψ_1 and Ψ_2 on the elementary sets of this σ-algebra. This gives a measure Υ_n which decreases to Υ as $n \to \infty$. The elementary sets generated by $B_1, \ldots, B_n$ are just the $\langle\sigma\rangle$, so that

$$\Upsilon_n(S) = \sum_{\sigma} \min[\Psi_1\langle\sigma\rangle, \Psi_2\langle\sigma\rangle].$$

Thus (8.35) implies that

$$\mathbb{E}\{w_n \Phi(S)\} \leqslant \mathbb{E}\{\Upsilon_n(S)\}, \tag{8.37}$$

and the right-hand side decreases with n to the limit $\mathbb{E}\{\Upsilon(S)\}$.

However, Ψ_1 and Ψ_2 arise from independent Poisson processes Π_1^* and Π_2^*, and their sets of atoms form independent Poisson processes on S. By the Disjointness Lemma, these sets are disjoint with probability 1, so that no non-zero measure lies below both Ψ_1 and Ψ_2. Here $\Upsilon(S) = 0$ almost certainly, and (8.37) shows that

$$\mathbb{E}\{w_n(S)\} \to 0. \tag{8.38}$$

The only question is whether this convergence is fast enough to ensure (8.33).

The computations necessary to establish (8.33) seem difficult, and to require a further condition on the distributions of Φ. We can for instance use the inequality

$$\begin{aligned}\mathbb{E}\{\min[\Psi_1\langle\sigma\rangle, \Psi_2\langle\sigma\rangle]\} &\leqslant \mathbb{E}\{\Psi_1\langle\sigma\rangle^{1/2}\Psi_2\langle\sigma\rangle^{1/2}\} \\ &= \{\mathbb{E}[\Psi\langle\sigma\rangle^{1/2}]\}^2\end{aligned}$$

and the estimate

$$\begin{aligned}\mathbb{E}[\Psi\langle\sigma\rangle^{1/2}\} &= \mathbb{E}[\textstyle\sum \{z; (x, z) \in \Pi^*, x \in \langle\sigma\rangle\}]^{1/2} \\ &\leqslant \mathbb{E}[\textstyle\sum \{z^{1/2}; (x, z) \in \Pi^*, x \in \langle\sigma\rangle\}] \\ &= \displaystyle\int z^{1/2}\gamma(\langle\sigma\rangle, \mathrm{d}z),\end{aligned}$$

so long as the measure

$$\rho(A) = \int_{(0,\infty)} z^{1/2}\gamma(A, \mathrm{d}z) \tag{8.39}$$

is finite on S. The left-hand side of (8.38) is then at most

$$\sum_{\sigma} \rho\langle\sigma\rangle^2, \tag{8.40}$$

the quadratic variation of ρ on $\mathscr{B}_n$. This always tends to zero, and will do so sufficiently fast if the B_n are well chosen. The best possible choice is one that divides ρ equally, so that $\rho\langle\sigma\rangle = \rho(S)/2^n$ for all σ of length n. The bound (8.40) is then equal to 2^{-n}, and the Blackwell argument succeeds.

It should be said that these sufficient conditions are probably far from necessary, and there may well be much more powerful ways of proving that large families of completely random measures (with $\beta = 0$) are purely atomic with probability 1.

8.4 Subordinators

We now specialise the theory to the one-dimensional case. A completely random measure Φ on $\mathbb{R}$, finite on bounded sets, defines a random process in the usual sense, a random function ϕ: $\mathbb{R} \to \mathbb{R}$, defined by

$$\begin{aligned}\phi(t) &= \Phi(0, t] \qquad (t \geqslant 0) \\ &= -\Phi(t, 0] \quad (t < 0).\end{aligned} \tag{8.41}$$

Clearly ϕ is increasing and right-continuous, and it determines Φ uniquely. The independence of Φ on finite sets implies that ϕ has independent increments, in the sense that, for $t_1 < t_2 < \cdots < t_n$, the increments

$$\phi(t_{r+1}) - \phi(t_r) \quad (r = 1, 2, \ldots, n-1), \tag{8.42}$$

are independent random variables. They are also positive, so that the theory of completely random measures on $\mathbb{R}$ (finite on bounded sets) is coterminous with the theory of random processes with positive independent increments.

In particular, we can say at once that such a process is the sum of three such processes. One is a deterministic increasing function, the second a random function that increases only at certain fixed discontinuities, while the third (and interesting) is derived from a Poisson process on the half-plane $S^* = \{(x, z); z > 0\}$ by the formula

$$\begin{aligned} \phi(t) &= \sum \{z; (x, z) \in \Pi^*, 0 < x \leqslant t\} \quad (t > 0) \\ &= -\sum \{z; (x, z) \in \Pi^*, t < x \leqslant 0\} \quad (t < 0). \end{aligned} \tag{8.43}$$

By far the most important case is that in which the distribution of a typical increment

$$\phi(t) - \phi(s) \quad (s < t)$$

depends only on the difference $t - s$. A process with positive independent increments having this property is called a *subordinator*, and such processes occur in many areas of pure and applied probability.

A subordinator cannot have fixed discontinuities, because the discontinuity set is countable and cannot therefore be invariant under translations. Similarly the deterministic component can only be a constant multiple of t. Thus a subordinator has, in $t \geqslant 0$, the representation

$$\phi(t) = \beta t + \Sigma \{z; (x, z) \in \Pi^*, 0 < x \leqslant t\}, \tag{8.44}$$

where β is a constant, and Π^* is a Poisson process on S^* which must now be invariant under translations parallel to the x-axis. The mean measure μ^* of Π^* must therefore be of the form

$$\mu^*(\mathrm{d}x\,\mathrm{d}z) = \mathrm{d}x\,\gamma(\mathrm{d}z), \tag{8.45}$$

where γ is a measure on $(0, \infty)$ such that

$$\int (1 - \mathrm{e}^{-z})\gamma(\mathrm{d}z) < \infty. \tag{8.46}$$

In particular, (8.9) shows that

$$\mathbb{E}\{\mathrm{e}^{-\theta[\phi(t) - \phi(s)]}\} = \exp\left\{-t\left[\beta\theta + \int_{(0,\infty)} (1 - \mathrm{e}^{-\theta z})\gamma(\mathrm{d}z)\right]\right\}. \tag{8.47}$$

The atoms of Φ correspond to jump discontinuities of ϕ. These occur at the points of the projection Π of Π^* onto the x-axis, which form (by the Mapping Theorem) a Poisson process of constant rate $\gamma(0, \infty)$. Notice however that (8.46) allows the possibility that $\gamma(0, \infty)$ is infinite, in which case the jumps of Π form a dense set. If however we consider only jumps of height greater than some $\delta > 0$, these form a homogeneous Poisson process of finite rate $\gamma(\delta, \infty)$.

One of the most important problems in which subordinators occur is in the theory of Brownian motion and more general diffusion processes. Suppose for instance that

$$W(t) \quad (t \geqslant 0)$$

is a simple Brownian motion, a process with independent (positive and negative) increments for which $W(t) - W(s)$ has, for $s < t$, a normal distribution with mean 0 and variance $t - s$. We can and shall take a version which is a continuous function of t.

The zero set

$$\mathscr{Z} = \{t \geqslant 0; W(t) = W(0) = 0\} \tag{8.48}$$

is a random closed set of great complexity. It is with probability 1 uncountable but of zero measure, so that one can speak neither of the 'number of visits to 0' nor of 'the total time spent at 0'.

It turns out however that there is a process $L(t)$ which measures, in a generalised sense, the time which W spends 'near' 0 up to time t. As a function of t, L is increasing but increases only on $\mathscr{Z}$. More precisely, L is constant on each of the intervals which make up the complement of the closed $\mathscr{Z}$. These intervals can be regarded as the *excursions* of W from 0.

The function L is continuous and tends to ∞ as $t \to \infty$, so that it has a right-continuous inverse function ϕ. It turns out that ϕ is a subordinator, and that $\mathscr{Z}$ is the closure of its range. The excursions of W correspond to the jumps of ϕ. Since these in turn correspond to the points of Π^*, the whole structure of $\mathscr{Z}$ can be described in terms of ϕ and thus of Π^*.

The measure γ turns out to be given by

$$\gamma(\mathrm{d}z) = z^{-3/2}\,\mathrm{d}z, \tag{8.49}$$

up to a constant which is arbitrary in the definition of the *local time process* L. Note that $\gamma(0, \infty) = \infty$.

For the justification of the statements of the last few paragraphs, and for much else, see Williams (1979) or Itó and McKean (1965). The argument is not confined to Brownian motion, but applies (with different γ) to many continuous time Markov processes. The local time process can be recovered from $\mathscr{Z}$ and γ, in a simple way depending non-trivially on the strong law of Section 4.5 (Kingman 1973).

9

The Poisson–Dirichlet distribution

9.1 The Dirichlet distribution

It is often necessary to consider random vectors

$$p = (p_1, p_2, \ldots, p_n) \tag{9.1}$$

which form a discrete probability distribution, so that

$$p_j \geqslant 0 \quad (j = 1, 2, \ldots, n), \qquad \sum_{j=1}^{n} p_j = 1. \tag{9.2}$$

For example, p_j might be the proportion of a biological population of the jth of n possible types. Random probability vectors of this kind also arise in the Bayesian approach to statistical theory.

By far the simplest non-trivial example of a probability distribution over the simplex Δ_n of vectors satisfying (2) is the *Dirichlet distribution* $\mathscr{D}(\alpha_1, \alpha_2, \ldots, \alpha_n)$ whose density (relative to $(n-1)$-dimensional Lebesgue measure on Δ_n) is given by

$$f(p_1, p_2, \ldots, p_n) = \frac{\Gamma(\alpha_1 + \alpha_2 + \cdots + \alpha_n)}{\Gamma(\alpha_1)\Gamma(\alpha_2)\cdots\Gamma(\alpha_n)} p_1^{\alpha_1 - 1} p_2^{\alpha_2 - 1} \cdots p_n^{\alpha_n - 1}. \tag{9.3}$$

The parameters α_j may take any strictly positive values, and the character of the distribution changes markedly as these vary. If $\alpha_j = 1$ for all j we have the uniform distribution on Δ_n. If the α_j are large, (9.3) concentrates probability well away from the boundaries of Δ_n, corresponding to distributions p which are fairly evenly spread. On the other hand, small values of the α_j concentrate near the boundary of Δ_n, corresponding to highly disparate p with a few large p_j and the others small.

In particular, if all the α_j are equal to a small value α, the p_j have by symmetry the same expectation $1/n$, but there is high probability that at least one of the p_j is much greater than that average; which value or values of j have large p_j is a matter only of chance. In Section 9.3 we shall prove a limiting result which makes this precise.

Direct manipulation of (9.3) is difficult, because of the linear dependence of the p_j. It has long been known that a better way to handle the distribution is in terms of independent gamma variables. Let $Y_1, Y_2, \ldots, Y_n$ be independent positive random variables, Y_j having the probability density

$$g_\alpha(y) = y^{\alpha - 1}\, e^{-y}/\Gamma(\alpha) \quad (y > 0) \tag{9.4}$$

with $\alpha = \alpha_j$. Let $Y = Y_1 + Y_2 + \cdots + Y_n$; then it is easy to see that the vector p with components

$$p_j = Y_j/Y \tag{9.5}$$

has the distribution $\mathscr{D}(\alpha_1, \alpha_2, \ldots, \alpha_n)$ and is independent of Y. The proof is an immediate calculation by a change of variables according to the function from $\mathbb{R}^n$ to $\mathbb{R}^n$ given by

$$(Y_1, Y_2, \ldots, Y_n) \mapsto (Y, p_1, p_2, \ldots, p_{n-1});$$

the details are left to the reader.

It is a corollary of that calculation that Y also has distribution (9.4), with

$$\alpha = \alpha_1 + \alpha_2 + \cdots + \alpha_n.$$

This can also be proved using the Laplace transform

$$\int_0^\infty g_\alpha(y)\,\mathrm{e}^{-\theta y}\,\mathrm{d}y = \frac{1}{(1+\theta)^\alpha} \quad (\theta > -1) \tag{9.6}$$

which shows too that the distribution $\mathscr{G}(\alpha)$ given by (9.4) is infinitely divisible, with Lévy–Khinchin representation

$$\frac{1}{(1+\theta)^\alpha} = \exp\left\{-\alpha \int_0^\infty (1 - \mathrm{e}^{-\theta z}) z^{-1}\,\mathrm{e}^{-z}\,\mathrm{d}z\right\}. \tag{9.7}$$

Corresponding to (9.7) is a subordinator known as the *Moran gamma process*, which was used by Moran (1959) in his pioneering theory of the storage of water by dams. This process is defined exactly as in Section 8.4 with

$$\beta = 0, \qquad \gamma(\mathrm{d}z) = z^{-1}\,\mathrm{e}^{-z}\,\mathrm{d}z, \tag{9.8}$$

and is such that the increment $\phi(t) - \phi(s)$ has distribution $\mathscr{G}(t-s)$. Note that

$$\gamma(0, \infty) = \int_0^\infty z^{-1}\,\mathrm{e}^{-z}\,\mathrm{d}z = \infty,$$

so that the jumps of ϕ are everywhere dense.

For $\alpha_1, \alpha_2, \ldots, \alpha_n > 0$, define

$$t_0 = 0, \qquad t_j = \alpha_1 + \alpha_2 + \cdots + \alpha_j \quad (1 \leqslant j \leqslant n). \tag{9.9}$$

Then $Y_j = \phi(t_j) - \phi(t_{j-1})$ has distribution $\mathscr{G}(\alpha_j)$ and the Y_j are independent. Since

$$Y = Y_1 + Y_2 + \cdots + Y_n = \phi(t_n),$$

we see that (9.1) with

$$p_j = \{\phi(t_j) - \phi(t_{j-1})\}/\phi(t_n) \tag{9.10}$$

defines a random vector in Δ_n with distribution $\mathscr{D}(\alpha_1, \alpha_2, \ldots, \alpha_n)$. This representation will prove in Section 9.3 to be a powerful tool for the elucidation and manipulation of $\mathscr{D}(\alpha_1, \alpha_2, \ldots, \alpha_n)$ when n is large and the α_j are small.

9.2 The Dirichlet process

The Moran gamma process corresponds as in Section 8.4 to a completely random measure Φ on $\mathbb{R}$ having the property that $\Phi(A)$ has the gamma distribution $\mathscr{G}(\alpha)$ with α equal to the length of A. In this form the concept generalises at once to quite arbitrary spaces.

Let μ be a measure on the space S (arbitrary other than the conditions of Section 2.1), and apply the construction of Section 8.2 with

$$\gamma(A, B) = \mu(A) \int_B z^{-1} \mathrm{e}^{-z}\, \mathrm{d}z. \tag{9.11}$$

Thus the Poisson process Π^* on $S^* = S \times (0, \infty)$ has mean measure which is the product of μ and the measure (9.8), and

$$\Phi(A) = \Sigma\, \{z; (x, z) \in \Pi^*, x \in A\} \tag{9.12}$$

defines a completely random measure Φ with

$$\begin{aligned}\mathbb{E}\{\mathrm{e}^{-t\Phi(A)}\} &= \exp\left\{-\int_0^\infty \mu(A)(1 - \mathrm{e}^{-tz}) z^{-1} \mathrm{e}^{-z}\, \mathrm{d}z\right\}\\ &= (1 + t)^{-\mu(A)}.\end{aligned}$$

Thus $\Phi(A)$ has the gamma distribution $\mathscr{G}(\mu(A))$ for any A with $\mu(A)$ finite.

Let $A_1, A_2, \ldots, A_n$ be disjoint with union A, with

$$\alpha_j = \mu(A_j) < \infty. \tag{9.13}$$

Then the random variables

$$Y_j = \Phi(A_j) \tag{9.14}$$

are independent with distributions $\mathscr{G}(\alpha_j)$, and

$$Y = Y_1 + Y_2 + \cdots + Y_n = \Phi(A).$$

Hence (9.5) defines a random vector $p \in \Delta_n$ whose distribution is $\mathscr{D}(\alpha_1, \alpha_2, \ldots, \alpha_n)$.

In particular, if μ is totally finite, $\Phi(S)$ is finite with probability one, and

$$\Upsilon(A) = \Phi(A)/\Phi(S) \tag{9.15}$$

defines a random probability measure on S such that, for any partition $A_1, A_2, \ldots, A_n$ of S, the joint distribution of

$$\Upsilon(A_1), \Upsilon(A_2), \ldots, \Upsilon(A_n) \tag{9.16}$$

is $\mathscr{D}(\alpha_1, \alpha_2, \ldots, \alpha_n)$ with $\alpha_j = \mu(A_j)$. The elements of (9.16) are not of course independent, so that Υ is not a completely random measure. It is called a *Dirichlet process* corresponding to the measure μ.

The Blackwell argument described in Section 8.3 was originally designed for the Dirichlet process, or its near relative the completely random measure governed by (9.11). See Ferguson (1973) for a detailed account, and for applications to Bayesian statistics.

9.3 The Poisson–Dirichlet limit

After this digression into general spaces, we now return to the one-dimensional context of Section 9.1. Suppose that $p \in \Delta_n$ is a random vector with distribution $\mathscr{D}(\alpha_1, \alpha_2, \ldots, \alpha_n)$, where for definiteness we take the α_j to be equal, to α say. By symmetry,

$$\mathbb{E}(p_j) = 1/n, \tag{9.17}$$

so that when n is large each p_j is likely to be small.

This is true whatever the value of α, but when α is small the values of p_j are, with high probability, disparate in value. It may well happen that some of the p_j (though necessarily only a few) are not small.

Thus let

$$p_{(1)} \geqslant p_{(2)} \geqslant \cdots \geqslant p_{(n)} \tag{9.18}$$

denote the p_j arranged in descending order. Can anything be said about $p_{(1)}$, or more generally $p_{(k)}$, or more generally still the vector $(p_{(1)}, p_{(2)}, \ldots, p_{(k)})$ for fixed k as $n \to \infty$ and $\alpha \to 0$? The answer is yes, provided that $n\alpha$ converges to some finite non-zero limit λ.

More generally, suppose that (9.18) represents the elements, arranged in descending order, of a random vector $p^{(n)}$ with Dirichlet distribution

$$\mathscr{D}(\alpha_1^{(n)}, \alpha_2^{(n)}, \ldots, \alpha_n^{(n)}).$$

Suppose that, as $n \to \infty$,

$$\max(\alpha_1^{(n)}, \alpha_2^{(n)}, \ldots, \alpha_n^{(n)}) \to 0 \tag{9.19}$$

and that

$$\lambda^{(n)} = \alpha_1^{(n)} + \alpha_2^{(n)} + \cdots + \alpha_n^{(n)} \to \lambda. \tag{9.20}$$

Then there are limiting results for the $p_{(k)}^{(n)}$ for fixed k as $n \to \infty$.

To see what these must be, let ϕ be the gamma process of Section 9.1, and define

$$\bar{p}_j^{(n)} = \{\phi(\alpha_1^{(n)} + \cdots + \alpha_j^{(n)}) - \phi(\alpha_1^{(n)} + \cdots + \alpha_{j-1}^{(n)})\}/\phi(\lambda^{(n)}). \quad (9.21)$$

Then for each n the vector $\bar{p}^{(n)} = (\bar{p}_1^{(n)}, \ldots, \bar{p}_n^{(n)})$ has the same Dirichlet distribution as $p^{(n)}$, and the joint distribution of the $\bar{p}_{(k)}^{(n)}$ are the same as those of the $p_{(k)}^{(n)}$. But under the conditions (9.19) and (9.20) the $\bar{p}_{(k)}^{(n)}$ actually converge as $n \to \infty$, without any considerations of probability. Thus let $J_1 \geqslant J_2 \geqslant J_3 \geqslant \cdots$ be the heights of the jumps of ϕ in $[0, \lambda]$, arranged in descending order. Elementary real analysis applied to (9.22) shows that, as $n \to \infty$,

$$\phi(\lambda^{(n)})\bar{p}_{(k)}^{(n)} \to J_k. \quad (9.22)$$

It follows that, for each k, the joint distribution of

$$\bar{p}_{(1)}^{(n)}, \bar{p}_{(2)}^{(n)}, \ldots, \bar{p}_{(k)}^{(n)}$$

converges to that of $\xi_1, \xi_2, \ldots, \xi_k$, where

$$\xi_k = J_k/\phi(\lambda), \quad (9.23)$$

and the same is therefore true of

$$p_{(1)}^{(n)}, p_{(2)}^{(n)}, \ldots, p_{(k)}^{(n)}.$$

Since ϕ increases only in jumps,

$$\phi(\lambda) = \sum_{k=1}^{\infty} J_k, \quad (9.24)$$

so that

$$\xi_1 \geqslant \xi_2 \geqslant \cdots, \quad \sum_{k=1}^{\infty} \xi_k = 1. \quad (9.25)$$

The distribution of the infinite random sequence $\xi = (\xi_1, \xi_2, \ldots)$ satisfying (9.25), depends only on λ and is called the *Poisson–Dirichlet distribution* $\mathscr{PD}(\lambda)$. Thus we can summarise the result by saying that the decreasing order statistics of the Dirichlet distribution $\mathscr{D}(\alpha_1, \alpha_2, \ldots, \alpha_n)$ approximate to those of $\mathscr{PD}(\lambda)$, so long as n is large, the α_j are uniformly small, and $\alpha_1 + \alpha_2 + \cdots + \alpha_n$ is near λ.

This turns out to be of great importance in a variety of applications, notably in population genetics and ecology. The Dirichlet distribution is the equilibrium distribution for a variety of evolutionary models, and small α_j with large n are a typical combination. The reader may wish to consult Kingman (1980).

9.4 The Moran subordinator

The Poisson–Dirichlet distribution is the distribution of an infinite sequence $\xi = (\xi_1, \xi_2, \xi_3, \ldots)$ satisfying

$$\xi_1 \geqslant \xi_2 \geqslant \xi_3 \geqslant \cdots, \qquad \sum_{k=1}^{\infty} \xi_k = 1. \tag{9.26}$$

Such a sequence with this distribution can be generated by the formula

$$\xi_k = J_k(\lambda)/\phi(\lambda), \tag{9.27}$$

where $J_k(\lambda)$ is the height of the kth largest jump in $[0, \lambda]$ of the Moran process ϕ. Now the positions and heights of these jumps form a Poisson process Π^* in the plane with mean measure

$$\mu^*(\mathrm{d}x\,\mathrm{d}z) = z^{-1}\,\mathrm{e}^{-z}\,\mathrm{d}x\,\mathrm{d}z \quad (z > 0). \tag{9.28}$$

Hence, by the Mapping Theorem, the heights of the jumps in $[0, \lambda]$ form a Poisson process Π_λ on $(0, \infty)$ with rate

$$\lambda z^{-1}\,\mathrm{e}^{-z}. \tag{9.29}$$

Because this rate function is not integrable up to $z = 0$, there are infinitely many points, with 0 as a limit point. The $J_k(\lambda)$ are the points of Π_λ arranged in descending order, and

$$\phi(\lambda) = \sum_{k=1}^{\infty} J_k(\lambda) < \infty. \tag{9.30}$$

The argument reverses to give a direct description of $\mathscr{PD}(\lambda)$ in terms of a Poisson process Π_λ with rate function (9.29). If $\eta_1 \geqslant \eta_2 \geqslant \eta_3 \geqslant \cdots$ are the points of such a process, then (Campbell's Theorem)

$$\sigma = \sum_{k=1}^{\infty} \eta_k < \infty, \tag{9.31}$$

and

$$\xi_k = \eta_k/\sigma \tag{9.32}$$

defines a sequence ξ having distribution $\mathscr{PD}(\lambda)$. It is useful for calculations to note that ξ and σ are independent, this being a direct consequence of the independence of p and Y in Section 9.1 (as well as being easy to prove directly).

Equation (9.31) shows that η_k must tend to zero fairly fast, but a more precise result comes from the strong law of Section 4.5. Applying (4.59) to the process Π_λ we have

$$\lim_{t \to \infty} \#\{k; \eta_k > t\}/L(t) = 1 \tag{9.33}$$

with probability 1, where

$$L(t) = \int_t^{\infty} \lambda z^{-1}\,\mathrm{e}^{-z}\,\mathrm{d}z \sim -\log t \tag{9.34}$$

as $t \to 0$. Thus

$$\#\{k; \eta_k > t\} \sim -\lambda \log t$$

as $t \to 0$, which implies that

$$-\log \eta_k \sim k/\lambda \tag{9.35}$$

as $k \to \infty$, with probability 1. Thus η_k decays exponentially fast, as does ξ_k; by (9.32)

$$-\log \xi_k \sim k/\lambda. \tag{9.36}$$

9.5 The Ewens sampling formula

Because of its derivation from the Dirichlet distribution, the distribution $\mathscr{PD}(\lambda)$ arises frequently as a model for the division of a large population between a large number of possible species or types. The infinite random vector $\xi = (\xi_1, \xi_2, \ldots)$ with distribution $\mathscr{PD}(\lambda)$ describes the structure of the population, ξ_k being the proportion of the kth most common type.

To illustrate its use, suppose that from such a large population a random sample of size n is taken. What is the probability that all the elements of the sample are of the same type? Conditional on ξ this is $\sum_{k=1}^{\infty} \xi_k^n$, so that the unconditional probability is

$$h_n = \mathbb{E}\left\{\sum_{k=1}^{\infty} \xi_k^n\right\}. \tag{9.37}$$

By (9.32) and the independence of ξ and σ,

$$\begin{aligned}\mathbb{E}\left\{\sum_{k=1}^{\infty} \eta_k^n\right\} &= \mathbb{E}\left\{\sigma^n \sum_{k=1}^{\infty} \xi_k^n\right\} \\ &= \mathbb{E}\{\sigma^n\} h_n.\end{aligned}$$

Now σ has distribution $\mathscr{G}(\lambda)$ and so

$$\mathbb{E}\{\sigma^n\} = \Gamma(n+\lambda)/\Gamma(\lambda), \tag{9.38}$$

and Campbell's Theorem gives

$$\mathbb{E}\left\{\sum_{k=1}^{\infty} \eta_k^n\right\} = \int_0^{\infty} z^n \lambda z^{-1} \mathrm{e}^{-z}\,\mathrm{d}z = \lambda(n-1)!.$$

Thus

$$h_n = \frac{\lambda\Gamma(\lambda)(n-1)!}{\Gamma(n+\lambda)} = \frac{(n-1)!}{(1+\lambda)(2+\lambda)\cdots(n+\lambda-1)}. \tag{9.39}$$

This is a very special case of a justly famed result known as the *Ewens sampling formula*. If the sample of size n is not homogeneous, there may be some unique members, of the same type as no others in the sample. Say there are a_1 of them. Say also that there are a_2 pairs, each pair of the same type but different from all others, a_3 triples, a_4 quartets and so on. The numbers $a_1, a_2, \ldots$ satisfying

$$a_1, a_2, \ldots, a_n \geqslant 0, \qquad a_1 + 2a_2 + \cdots + na_n = n, \tag{9.40}$$

describe the sample as well as may be in the absence of any labelling of the different types. They can be summed up as a partition

$$\mathfrak{a} = 1^{a_1} 2^{a_2} \cdots n^{a_n} \tag{9.41}$$

of the sample size n, and $\mathfrak{a}$ is a random partition whose distribution under $\mathscr{PD}(\lambda)$ can be calculated. If $P_n(\mathfrak{a})$ denotes the probability of the sample exhibiting the partition $\mathfrak{a}$, then (9.39) gives the particular case $P_n(n^1)$.

The Ewens formula gives a corresponding expression for the general partition $\mathfrak{a}$.

To calculate $P_n(\mathfrak{a})$, note that the probability of achieving $\mathfrak{a}$ given the population frequencies ξ_k (which will now be written $\xi(k)$ for convenience) is

$$\frac{n!}{\Pi(j!)^{a_j} a_j!} \sum \xi(k_{11})\xi(k_{12}) \cdots \xi(k_{1a_1})\xi(k_{21})^2 \cdots \xi(k_{2a_2})^2 \xi(k_{31})^3 \cdots,$$

where the summation ranges over distinct

$$k_{ij} \quad (i = 1, 2, \ldots, n; \quad j = 1, 2, \ldots, a_i).$$

Hence

$$\frac{\Gamma(n + \lambda)}{\Gamma(\lambda)} P_n(\mathfrak{a}) = \frac{n!}{\Pi(j!)^{a_j} a_j!} \mathbb{E}\left\{\sum \eta(k_{11}) \cdots \eta(k_{1a_1})\eta(k_{21})^2 \cdots\right\}.$$

The evaluation of expected sums of this sort was explained in Section 3.2; equation (3.28) shows that the expectation on the right equals

$$\prod_{j=1}^{n} \mathbb{E}\left\{\sum_{k=1}^{\infty} \eta(k)^j\right\}^{a_j} = \prod_{j=1}^{n} \left\{\int_0^{\infty} z^j \lambda z^{-1} \mathrm{e}^{-z} \, \mathrm{d}z\right\}^{a_j}$$

$$= \prod_{j=1}^{n} \{\lambda (j-1)!\}^{a_j}.$$

Substituting and simplifying, we have finally

$$P_n(\mathfrak{a}) = \frac{n! \, \Gamma(\lambda)}{\Gamma(n + \lambda)} \prod_{j=1}^{n} \left(\frac{\lambda^{a_j}}{j^{a_j} a_j!}\right). \tag{9.42}$$

This formula has been established for many different models since it was first propounded by Ewens (1972); see Kingman (1980) for details. In other contexts it is older, since Cauchy showed that the cycle partition of a random permutation has distribution (9.42) with $\lambda = 1$.

Note that (9.42) can be written in terms of Poisson probabilities:

$$P_n(\mathfrak{a}) = C(n, \lambda) \prod_{j=1}^{n} \pi_{a_j}(\lambda/j). \tag{9.43}$$

Thus the joint distribution of the a_j is the same as if they were independent with respective distributions $\mathscr{P}(\lambda/j)$, but conditioned by the constraint (9.40).

The conditioning makes calculations difficult, but its force diminishes as n increases. Thus it has been shown by Arratia, Barbour and Tavaré (1992) that, if random variables $Z_1(n), Z_2(n), \ldots, Z_n(n)$ have the joint distribution (9.42), then as $n \to \infty$ the joint distribution of $Z_1(n), Z_2(n), \ldots, Z_N(n)$ for fixed N converges to that of $\zeta_1, \zeta_2, \ldots, \zeta_N$ where the ζ_j are independent with respective distributions $\mathscr{P}(\lambda/j)$.

9.6 Size-biased sampling

The calculations of the last section show that the distribution $\mathscr{P}\mathscr{D}(\lambda)$ is rather less than user-friendly. It is therefore of some interest to note another distribution, of simpler form, from which $\mathscr{P}\mathscr{D}(\lambda)$ can be derived. This goes back to work of Patil and Taillie (1977; see also Donnelly 1986).

Let the random probability vector

$$p = (p_1, p_2, \ldots, p_n) \tag{9.44}$$

have the symmetric Dirichlet distribution $\mathscr{D}(\alpha, \alpha, \ldots, \alpha\}$.

Let v be a random variable having values $1, 2, \ldots, n$ in such a way that

$$\mathbb{P}\{v = r | p\} = p_r \quad (1 \leqslant r \leqslant \alpha). \tag{9.45}$$

Then a standard calculation shows that the vector

$$p' = (p_v, p_1, \ldots, p_{v-1}, p_{v+1}, \ldots, p_n) \tag{9.46}$$

has distribution

$$\mathscr{D}(\alpha + 1, \alpha, \ldots, \alpha). \tag{9.47}$$

It follows that $(p_v, 1 - p_v)$ has the Dirichlet distribution

$$\mathscr{D}(\alpha + 1, (n - 1)\alpha)$$

so that p_v has probability density

$$\frac{\Gamma(n\alpha + 1)}{\Gamma(\alpha + 1)\Gamma(n\alpha - \alpha)} p^{\alpha}(1 - p)^{(n-1)\alpha - 1}. \tag{9.48}$$

Given p_v, the conditional joint distribution of the remaining components of p is the same as that of $(1 - p_v)p^{(1)}$, where the $(n - 1)$-vector $p^{(1)}$ has the symmetric distribution $\mathscr{D}(\alpha, \alpha, \ldots, \alpha)$.

We say that p_ν is *obtained from p by size-biased sampling.* This process may now be applied to $p^{(1)}$ to produce a component with distribution (9.48) (but with n replaced by $n-1$) and an $(n-2)$-vector $p^{(2)}$ with distribution $\mathscr{D}(\alpha, \alpha, \ldots, \alpha)$. Continuing in this way, we can see that the components of p can be rearranged as a vector q whose components may be written

$$q_1 = v_1, \qquad q_2 = (1 - v_1)v_2, \qquad q_3 = (1 - v_1)(1 - v_2)v_3, \ldots. \tag{9.49}$$

The random variables $v_1, v_2, \ldots, v_{n-1}$ are independent, and v_r has density (9.48) with n replaced by $n - r + 1$.

Now let $\alpha \to 0$ and $n \to \infty$ with $n\alpha \to \lambda$. In this limit (9.48) converges to

$$\lambda(1 - p)^{\lambda - 1}. \tag{9.50}$$

Thus, if $q_1, q_2, q_3, \ldots$ are *defined* by (9.49), where the v_r are independent with density (9.50), and if p_k is the kth largest of the q_j then the sequence

$$(p_1, p_2, p_3, \ldots) \tag{9.51}$$

has distribution $\mathscr{PD}(\lambda)$.

References

Arratia, R., Barbour, A. D., and Tavaré, S. (1992). Poisson process approximations for the Ewens sampling formula. *Advances in Applied Probability*. (In press.)

Asmussen, S. (1987). *Applied probability and queues*. Wiley, New York.

Bartlett, M. S. (1949). Some evolutionary stochastic processes. *Journal of the Royal Statistical Society* B **11**, 211–29.

Billingsley, P. (1979). *Probability and measure*. Wiley, New York.

Blackwell, D. (1973). Discreteness of Ferguson selections. *Annals of Statistics* **1**, 356–8.

Breiman, L. (1968). *Probability*. Addison-Wesley, Reading, MA.

Campbell, N. R. (1909). The study of discontinuous phenomena. *Proceedings of the Cambridge Philosophical Society* **15**, 117–36.

Campbell, N. R. (1910). Discontinuities in light emission. *Proceedings of the Cambridge Philosophical Society* **15**, 310–28.

Chung, K. L. (1968). *A course in probability theory*. Harcourt, Brace & World, New York.

Cox, D. R. (1955). Some statistical models related with series of events. *Journal of the Royal Statistical Society* B **17**, 129–64.

Cox, D. R. (1962). *Renewal theory*. Methuen, London.

Cox, D. R. and Smith, W. L. (1961). *Queues*. Methuen, London.

Donnelly, P. (1986). Partition structures, Pólya urns, the Ewens sampling formula, and the age of alleles. *Theoretical Population Biology* **30**, 271–88.

Doob, J. L. (1953). *Stochastic processes*. Wiley, New York.

Ewens, W. J. (1972). The sampling theory of selectively neutral alleles. *Theoretical Population Biology* **3**, 87–112.

Felsenstein, J. (1975). A pain in the torus: some difficulties with models of isolation by distance. *The American Naturalist* **109**, 359–68.

Ferguson, T. S. (1973). A Bayesian analysis of some nonparametric problems. *Annals of Statistics* **1**, 209–30.

Grandell, J. (1976). *Doubly stochastic Poisson processes*. Springer-Verlag, Berlin.

Halmos, P. (1950). *Measure theory*. Van Nostrand, Princeton, NJ.

Harding, E. F. and Kendall, D. G. (ed.) (1974). *Stochastic geometry*. Wiley, New York.

Harris, T. E. (1963). *The theory of branching processes*. Springer-Verlag, Berlin.

Ito, K. and McKean, H. P. (1965). *Diffusion processes and their sample paths*. Springer-Verlag, Berlin.

Kallenberg, O. (1977). A counterexample to R. Davidson's conjecture on line processes. *Mathematical Proceedings of the Cambridge Philosophical Society* **82**, 301–7.

Kendall, D. G. (1974). Foundations of a theory of random sets. In *Stochastic geometry*, ed. E. F. Harding and D. G. Kendall, Wiley, New York.

Kendall, M. G. and Moran, P. A. P. (1963). *Geometric probability*. Griffin, London.

Kingman, J. F. C. (1964). On doubly stochastic Poisson processes. *Proceedings of the Cambridge Philosophical Society* **60**, 923–30.

Kingman, J. F. C. (1972). *Regenerative phenomena.* Wiley, New York.
Kingman, J. F. C. (1973). An intrinsic description of local time. *Journal of the London Mathematical Society* **6**, 725–31.
Kingman, J. F. C. (1980). *Mathematics of genetic diversity.* Society for Industrial and Applied Mathematics, Philadelphia, PA.
Kingman, J. F. C. and Taylor, S. J. (1966). *Introduction to measure and probability.* Cambridge University Press.
Miles, R. E. (1969, 1971). Poisson flats in Euclidean space. *Advances in Applied Probability* **1**, 211–37; **3**, 1–43.
Moran, P. A. P. (1959). *The theory of storage.* Methuen, London.
Moran, P. A. P. (1967). A non-Markovian quasi-Poisson process. *Studia Scientiarum Mathematicarum Hungarica* **2**, 425–9.
Nachbin, L. (1965). *The Haar integral.* Van Nostrand, Princeton, NJ.
Patil, G. P. and Taillie, C. (1977). Diversity as a concept and its implications for random environments. *Bulletin of the International Statistical Institute* **47**, 497–515.
Poisson, S.-D. (1837). *Recherches sur la probabilité des jugements en matière criminelle et en matière civile.* Bachelier, Paris.
Rényi, A. (1967). Remarks on the Poisson process. *Studia Scientiarum Mathematicarum Hungarica* **2**, 119–23.
Rényi, A. (1970). *Foundations of probability.* Holden-Day, San Francisco, CA.
Santaló, L. A. (1976). *Integral geometry and geometric probability.* Addison-Wesley, Reading, MA.
Takács, L. (1967). *Combinatorial methods in the theory of stochastic processes.* Wiley, New York.
Tanner, J. C. (1953). A problem of interference between two queues. *Biometrika* **40**, 58–69.
Whittaker, E. T. and Watson, G. N. (1902). *A course of modern analysis.* Cambridge University Press.
Williams, D. (1979). *Diffusions, Markov processes and martingales.* Wiley, New York.

Index of Theorems

Index

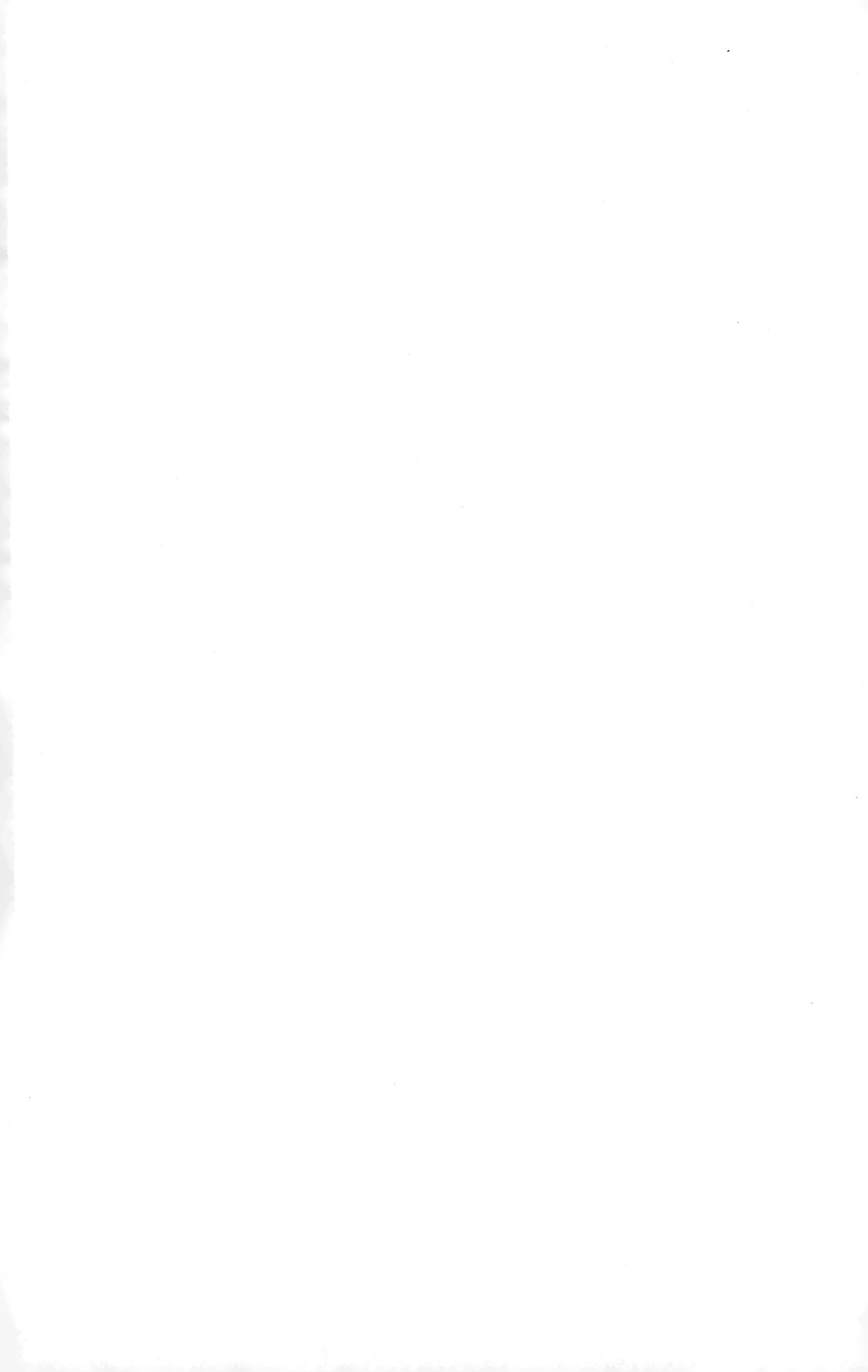